Lecture Notes in Computer Sc

T0238212

Commenced Publication in 1973
Founding and Former Series Editors:
Gerhard Goos, Juris Hartmanis, and Jan van Leeuwen

José Luiz Fiadeiro
Pierre-Yves Schobbens (Eds.)

Recent Trends in Algebraic Development Techniques

18th International Workshop, WADT 2006
La Roche en Ardenne, Belgium, June 1-3, 2006
Revised Selected Papers

 Springer

Volume Editors

José Luiz Fiadeiro
University of Leicester
Department of Computer Science
University Road, Leicester LE1 7RH, UK
E-mail: jose@mcs.le.ac.uk

Pierre-Yves Schobbens
Facultés Universitaires Notre-Dame de la Paix
Institut d'Informatique
Rue Grandgagnage 21, 5000 Namur, Belgium
E-mail: pys@info.fundp.ac.be

Library of Congress Control Number: 2007924494

CR Subject Classification (1998): F.3.1, F.4, D.2.1, I.1

LNCS Sublibrary: SL 1 – Theoretical Computer Science and General Issues

ISSN   0302-9743
ISBN-10   3-540-71997-0 Springer Berlin Heidelberg New York
ISBN-13   978-3-540-71997-7 Springer Berlin Heidelberg New York

Springer is a part of Springer Science+Business Media

springer.com

© Springer-Verlag Berlin Heidelberg 2007
Printed in Germany

Typesetting: Camera-ready by author, data conversion by Scientific Publishing Services, Chennai, India
Printed on acid-free paper   SPIN: 12049424   06/3180   5 4 3 2 1 0

Preface

This volume contains selected papers from WADT 2006, the 18th International Workshop on Algebraic Development Techniques. Like its predecessors, WADT 2006 focussed on the algebraic approach to the specification and development of systems, an area that was born around the algebraic specification of abstract data types and encompasses today the formal design of software systems, new specification frameworks and a wide range of application areas.

WADT 2006 took place at Chateau Floréal, La-Roche-en-Ardenne, Belgium, June 1–3, 2006, and was organized by Pierre-Yves Schobbens.

The program consisted of invited talks by David Rosenblum (University College London, UK) and Hubert Comon-Lundh (ENS-Cachan, France), and 32 presentations describing ongoing research on main topics of the workshop: formal methods for system development, specification languages and methods, systems and techniques for reasoning about specifications, specification development systems, methods and techniques for concurrent, distributed and mobile systems, and algebraic and co-algebraic foundations.

The Steering Committee of WADT, consisting of Michel Bidoit, José Fiadeiro, Hans-Jörg Kreowski, Till Mossakowski, Peter Mosses, Fernando Orejas, Francesco Parisi-Presicce, and Andrzej Tarlecki, with the additional help of Pierre-Yves Schobbens and Martin Wirsing, selected several presentations and invited their authors to submit a full paper for possible inclusion in this volume. All submissions underwent a careful refereeing process. We are extremely grateful to the following additional referees for their help in reviewing the submissions: A. Borzyszkowski, F. Gadducci, G. Godoy, K. Hölscher, A. Kurz, S. Kuske, A. Lopes, W. Pawlowski, H. Reichel, U. Schmid, L. Schröder, M. Sebag, and H. Wiklicky.

This volume contains the final versions of the ten contributions that were accepted.

The workshop was jointly organized with IFIP WG 1.3 (Foundations of System Specification), and received generous sponsorship from the University of Namur (Facultés Universitaires Notre-Dame de la Paix).

January 2007

José Fiadeiro
Pierre-Yves Schobbens

Table of Contents

Contributed Papers

A Temporal Graph Logic for Verification of Graph Transformation Systems*

Paolo Baldan[1], Andrea Corradini[2], Barbara König[3],
and Alberto Lluch Lafuente[2]

[1] Dipartimento di Matematica Pura e Applicata, Università di Padova
baldan@math.unipd.it
[2] Dipartimento di Informatica, Università di Pisa
{andrea,lafuente}@di.unipi.it
[3] Abt. für Informatik und Ang. Kognitionswissenschaft, Universität Duisburg-Essen
barbara_koenig@uni-due.de

Abstract. We extend our approach for verifying properties of graph transformation systems using suitable abstractions. In the original approach properties are specified as formulae of a propositional temporal logic whose atomic predicates are monadic second-order graph formulae. We generalize this aspect by considering more expressive logics, where edge quantifiers and temporal modalities can be interleaved, a feature which allows, e.g., to trace the history of objects in time. After characterizing fragments of the logic which can be safely checked on the approximations, we show how the verification of the logic over graph transformation systems can be reduced to the verification of a logic over suitably defined Petri nets.

1 Introduction

Graph Transformation Systems (GTSs) are suitable modeling formalisms for systems involving aspects such as object-orientation, concurrency, mobility and distribution. The use of GTSs for the verification and analysis of such systems is still at an early stage, but there have been several proposals recently, either using existing model-checking tools [10,25] or developing new techniques [20,21]. A recent line of research [1,2,3,4,5] takes the latter approach and proposes a method inspired by abstract interpretation techniques. Roughly speaking, a GTS $\mathcal{R}$, whose state space might be infinite, is approximated by a finite structure $\mathcal{C}(\mathcal{R})$, called *covering* of $\mathcal{R}$. The covering is a Petri net-like structure, called Petri graph, and it approximates $\mathcal{R}$ in the sense that any graph G reachable in $\mathcal{R}$ has an homomorphic image reachable in $\mathcal{C}(\mathcal{R})$. In a sense, this reduces the verification of GTSs to the verification of Petri nets. One central feature of this approach is

---

* Research partially supported by the EU IST-2004-16004 SENSORIA, the MIUR PRIN 2005015824 ART, the DFG project SANDS and CRUI/DAAD VIGONI "Models based on Graph Transformation Systems: Analysis and Verification".

J.L. Fiadeiro and P.-Y. Schobbens (Eds.): WADT 2006, LNCS 4409, pp. 1–20, 2007.
© Springer-Verlag Berlin Heidelberg 2007

the fact that it is a partial order reduction technique using unfoldings. That is, the interleaving of concurrent events—leading to state explosion—is avoided if possible.

In [5] a logic for the approximation approach is introduced, which is basically a propositional μ-calculus whose atomic predicates are closed formulae in a monadic second-order logic for graphs. Also, fragments of the logic are identified which are reflected by the approximations, i.e., classes of formulae which, when satisfied by the approximation, are satisfied by the original system as well. For the verification of such formulae, the logic is encoded into a μ-calculus whose atomic predicates are formulae over markings of a Petri net, allowing the reuse of existing model checking techniques for Petri nets [12].

There are other related papers working with graph logics, for instance [14]. However, most of them are based, like [5], on *propositional* temporal logics, that is, logics that do not allow to interleave temporal modalities with graph-related ones. Thus, properties like *a certain edge is never removed* are neither expressible nor verifiable. The only exceptions we are aware of are [20,22].

In this paper we extend the approach of [5] by considering a more expressive logic that allows to interleave temporal and graphical aspects. As we shall see, our temporal graph logic combines a monadic-second order logic of graphs with the μ-calculus. Formulae of our logic are interpreted over *graph transition systems* (GTrS), inspired by algebra transition systems [15] and the formalism of [20], which are traditional transition systems together with a function mapping states into graphs and transitions into partial graph morphisms. Graph transition systems are suitable formalisms for modeling the state space of graph transformation systems and Petri graphs. We introduce a notion of approximation for GTrSs, identifying fragments of the logic whose formulae are preserved or reflected by approximations. Then we show that the GTrS of the covering, as defined in [1], is an approximation of the GTrS of the original graph transformation system, thus providing a concrete way of constructing approximations. Finally, we propose an encoding for a fragment of our logic into a Petri net logic. Our encoding is correct and complete, i.e., a Petri graph satisfies a formula exactly when the encoding of the formula is satisfied by the underlying Petri net.

Putting all this together, given a graph transformation system $\mathcal{G}$ and a formula F in a suitable fragment of our logic, we can construct a Petri graph P which approximates $\mathcal{G}$, using the algorithm in [1]. Then F can be translated into a Petri net formula $[F]$, such that if N_P is the Petri net underlying P, then $N_P \models [F]$ implies $\mathcal{G} \models F$, i.e., we reduce verification over graph transformation systems to verification over Petri nets.

Section 2 introduces graphs, graph transformation systems and graph transition systems. Section 3 defines syntax and semantics of our temporal graph logic. Section 4 defines Petri graphs, the structures used for approximating graph transformation systems. Section 5 identifies fragments of the logic that are preserved or reflected by approximations. Section 6 proposes an encoding of a fragment of the logic into a Petri net logic. A last section concludes the paper and proposes further work.

2 Graph Transition Systems

An (edge-labeled) *graph* G is a tuple $G = \langle V_G, E_G, s_G, t_G, lab_G \rangle$ where V_G is a set of *nodes*, E_G is a set of *edges*, $s_G, t_G : E_G \to V_G$ are the *source* and *target* functions, and $lab_G : E_G \to \Lambda$ is a *labeling function*, where Λ is a fixed set of labels. Nodes and edges are sometimes called *items* and we shall write $X_G = E_G \cup V_G$ for the set of items of G.

The transformation of a graph into another by adding, removing or merging of items is suitably modeled by (partial) graph morphisms.

Definition 1 ((partial) graph morphism). *A graph morphism* $\psi : G_1 \to G_2$ *is a pair of mappings* $\psi_V : V_{G_1} \to V_{G_2}$, $\psi_E : E_{G_1} \to E_{G_2}$ *such that* $\psi_V \circ s_{G_1} = s_{G_2} \circ \psi_E$, $\psi_V \circ t_{G_1} = t_{G_2} \circ \psi_E$ *and* $lab_{G_1} = lab_{G_2} \circ \psi_E$. *A graph morphism* $\psi : G_1 \to G_2$ *is* injective *if so are* ψ_V *and* ψ_E; *it is* edge-injective *if* ψ_E *is injective.* Edge-bijective *morphisms are defined analogously. A graph* G' *is a* subgraph *of graph* G, *written* $G' \hookrightarrow G$, *if* $V_{G'} \subseteq V_G$ *and* $E_{G'} \subseteq E_G$, *and the inclusion is a graph morphism.*

A partial graph morphism $\psi : G_1 \rightharpoonup G_2$ *is a pair* $\langle G_1', \psi' \rangle$ *where* $G_1' \hookrightarrow G_1$ *is a subgraph of* G_1, *called the* domain *of* ψ, *and* $\psi' : G_1' \to G_2$ *is a graph morphism.*

Graph transformation is presented in set-theoretical terms, but could be equivalently presented by using the double-pushout [7] or single-pushout [11] approaches. With respect to more general definitions, our rules can neither delete nor merge nodes, and they have a discrete interface, i.e., the interface graph contains only nodes and thus edges are never preserved. Similar restrictions are assumed in [1]. While the condition of having a discrete interface can be relaxed, the deletion and merging of nodes is quite problematic in an unfolding-based approach.

Also observe that, as it commonly happens in the algebraic approaches to graph rewriting, we consider basic graph grammars, without any distinction between terminal and non-terminal symbols and without any high-level control mechanism. We remark that, even in this basic formulation, graph grammars are Turing complete (since they can simulate string rewriting).

Definition 2 (graph transformation system). *A graph transformation system* *(GTS)* $\mathcal{R}$ *is a pair* $\langle G_0, R \rangle$, *where* G_0 *is a start graph and* R *is a set of rewriting rules of the form* $r = \langle G_L, G_R, \alpha \rangle$, *where* G_L *and* G_R *are left- and right-hand side graphs, respectively, and* $\alpha : V_L \to V_R$ *is an injective function.*

A match *of a rule* r *in a graph* G *is a morphism* $\psi : G_L \to G$ *that is edge-injective. The application of a rule* r *to a match* ψ *in* G, *resulting in a new graph* H, *is called a* direct derivation *and is written* $G \xrightarrow{r, \psi} H$, *where* H *is defined as follows. The set* V_H *is* $V_G \uplus (V_R \setminus \alpha(V_L))$ *and* E_H *is* $(E_G \setminus \psi(E_L)) \uplus E_R$, *where* $\uplus$ *denotes disjoint union. The source, target and labeling functions are defined by*

$$s_H(e) = s_G(e) \quad t_H(e) = t_G(e) \quad lab_H(e) = lab_G(e) \qquad \text{if } e \in (E_G \setminus \psi(E_L)),$$
$$s_H(e) = \overline{\psi}(s_R(e)) \; t_H(e) = \overline{\psi}(t_R(e)) \; lab_H(e) = lab_R(e) \qquad \text{if } e \in E_R,$$

where $\overline{\psi} : V_R \to V_H$ *satisfies* $\overline{\psi}(\alpha(v)) = \psi(v)$ *if* $v \in V_L$ *and* $\overline{\psi}(v) = v$, *otherwise.*

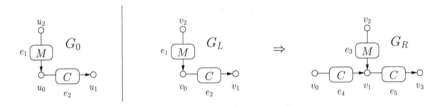

Fig. 1. A graph transformation system

Intuitively, the application of r to G at the match ψ first removes from G the image of the edges of L. Then the graph G is extended by adding the new nodes in G_R and the edges of G_R. All nodes in L are preserved.

A direct derivation $G \xrightarrow{r,\psi} H$ induces an obvious partial graph morphism $\tau_{G \xrightarrow{r,\psi} H} : G \rightarrow H$, injective and total on nodes, which maps items which are not deleted in G to corresponding items in H.

A *derivation* is a sequence of direct derivations starting from the start graph G_0. We write $G_0 \xrightarrow{*} H$ if there is a derivation ending in graph H, and we denote by $\mathbf{G}_{\mathcal{R}}$ the set of all graphs reachable in $\mathcal{R}$, i.e., $\mathbf{G}_{\mathcal{R}} = \{G \mid G_0 \xrightarrow{*} G\}$.

Example 1. Figure 1 depicts a GTS $\mathcal{G} = \langle G_0, \{r = \langle G_L, G_R, \alpha \rangle\}\rangle$ describing a simple message passing system. The start graph G_0 consists of three nodes u_0, u_1, u_2, one M-labeled edge e_1, representing a *message*, and one C-labeled edge e_2, representing a *connection*. The only rule consists of graphs G_L and G_R, and function α, which is the identity on V_L. The rule consumes the message and the connection, shifts the message to the successor node and recreates the connection. Furthermore a new C-labeled edge e_5 is created, along which the message can be passed in the next step. Note also that the source node of the message, representing its "identity", is preserved by the rule. In the rest of the paper we shall use this as a running example.

The state space of a GTS can be represented in a natural way as a transition system where the states are the reachable graphs and a transition between two states G and H exists whenever there is a direct derivation $G \xrightarrow{r,\psi} H$, as in [3]. However, such a structure would not be sufficient to interpret the logic introduced in the next section. Informally, since temporal modalities can be interleaved with quantification (over edges), the logic allows to trace the evolution of graph items over time and thus we need to represent explicitly which items of a graph are preserved by a rewriting step. To this aim, after recalling the standard definition of transition systems, we introduce below an enriched variant called *graph transition systems*.

Definition 3 (transition system). *A transition system is a tuple $M = \langle S_M,$ $T_M,\ in_M, out_M, s_0^M \rangle$ where S_M is a set of states, T_M is a set of transitions, $in_M, out_M : T_M \rightarrow S_M$ are functions mapping each transition to its start and end state, and $s_0^M \in S_M$ is the initial state. We shall sometimes write $s \xrightarrow{t} s'$*

if $in_M(t) = s$ *and* $out_M(t) = s'$, *and* $s \xrightarrow{*} s'$ *if there exists a (possibly empty) sequence of transitions from* s *to* s'.

Correspondingly, a transition system morphism $h : M \to M'$ is a pair of functions $\langle h^S : S_M \to S_{M'}, h^T : T_M \to T_{M'} \rangle$ such that the initial state as well as the start and end states of all transitions are preserved, i.e., $h^S(s_0^M) = s_0^{M'}$, $h^S \circ in_M = in_{M'} \circ h^T$, and $h^S \circ out_M = out_{M'} \circ h^T$.

A *graph transition system* is defined as a transition system together with a mapping which associates a graph with each state, and an injective partial graph morphism with each transition. We use the same name that is used in [20] for different, but closely related structures. The main difference is that in our case there is a clear distinction between the states and the graphs associated to the states: This leads below to a natural notion of morphism between graph transition systems, which will play a basic role in our definition of abstraction.

Definition 4 (graph transition system). *A* graph transition system *(GTrS)* $\mathcal{M}$ *is a pair* $\langle M, g \rangle$, *where* M *is a transition system and* g *is a pair* $g = \langle g^S, g^T \rangle$, *where* $g^S(s)$ *is a graph for each state* $s \in S_M$, *and* $g^T(t) : g^S(in_M(t)) \to g^S(out_M(t))$ *is an injective partial graph morphism for each transition* $t \in T_M$.

Note that the result of the application of a rule to a given match in a graph is determined only up to isomorphism, because of the use of disjoint union in the definition. Therefore, formally speaking, the state space of a GTS contains for each reachable graph G *all* graphs isomorphic to G as well. The next definition shows how to represent the state space of a GTS with a graph transition system (GTrS), where we get rid of such useless redundancy. Note that since the resulting GTrS is usually infinite-state, this construction is non-constructive and useless for practical purposes. We need the GTrS in order to define the semantics of the logic, but verification itself is done using a different method.

Definition 5 (graph transition system of a graph transformation system). *Given a GTS* $\mathcal{R} = \langle G_0, R \rangle$, *a GTrS representing its state space, denoted by* GTrS($\mathcal{R}$), *can be obtained as follows.*

1. *Consider first the graph transition system* $\langle M, g \rangle$, *where:* $S_M = \mathbf{G}_{\mathcal{R}}$ *(set of all graphs generated by* $\mathcal{R}$); $s_0^M = G_0$; $T_M = \{G \xrightarrow{r,\psi} H \mid G \xrightarrow{r,\psi} H$ *is a direct derivation of* $\mathcal{R}\}$; *and the mapping* $g = \langle g^S, g^T \rangle$ *is defined as follows:* $g^S(G) = G$ *and* $g^T(G \xrightarrow{r,\psi} H) = \tau_{G \xrightarrow{r,\psi} H} : G \to H$.
2. *Next, for each state* G *in* S_M *and for each pair* $\langle r, \psi \rangle$ *where* r *is a rule applicable to match* ψ *in* G, *choose one among the transitions leaving from* G *and using* r *and* ψ, *and delete from* T_M *all the remaining ones.*
3. *Finally,* GTrS($\mathcal{R}$) *is defined as the graph transition sub-system reachable from the start graph.*

Example 2. Figure 2 depicts a GTrS of the GTS depicted in Figure 1. Since state identities coincide with their corresponding graphs, the figure is simplified and we directly depict the graphs and partial graph morphisms. The leftmost state is

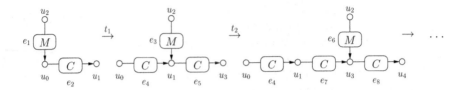

Fig. 2. A graph transition system

G_0, the initial state of both the GTS and its GTrS. Observe that for the second transition t_2, $g^T(t_2)_V$ (the component on nodes of $g^T(t_2)$) is an inclusion, while $g^T(t_2)_E$ is partial and is only defined on the edge e_4. All transitions correspond to different instances of the same rule.

The construction described in Definition 5 is clearly non-deterministic, because of step 2. Among the possible GTrSs associated with the GTS of Figure 1, the one drawn above enjoys some desirable properties: the partial injective morphisms associated with transitions are *partial inclusions* (i.e., every item preserves its name along rewriting), and edge and node names are not reused again in the computation after they have been deleted. The interpretation of the logic of Section 3 will be defined only over GTrSs satisfying such properties, and called *unraveled GTrSs*. For any GTrS $\mathcal{M}$ not satisfying these properties we shall consider an unraveled one which is behaviorally equivalent to $\mathcal{M}$, called its *unraveling*.

   In order to characterize the unraveling of a GTrS we first need to introduce GTrS morphisms, which will also be used later for relating a system and its approximation. A morphism between two GTrSs consists of a morphism between the underlying transition systems, and, in addition, for each pair of related states, of a graph morphism between the graphs associated with such states. Furthermore, these graph morphisms must be consistent with the partial graph morphisms associated to the transitions.

Definition 6 (graph transition system morphism). *A graph transition system morphism $h : \mathcal{M} \rightarrow \mathcal{M}'$ from $\mathcal{M} = \langle M, g \rangle$ to $\mathcal{M}' = \langle M', g' \rangle$ is a pair $\langle h_M, h_g \rangle$, where $h_M : M \rightarrow M'$ is a transition system morphism, and for each state $s \in S_M$, $h_g(s)$ is a graph morphism from $g^S(s)$ to $g'^S(h_M^S(s))$, such that the following condition is satisfied: for each transition $s_1 \xrightarrow{t} s_2 \in T_M$, $g'^T(h_M^T(t)) \circ h_g(s_1) = h_g(s_2) \circ g^T(t)$.*

The diagram below illustrates the situation. The bottom square represents transition $s_1 \xrightarrow{t} s_2$ in M and its image through h_M in M' (sub- and super-scripts are avoided for the sake of readability). The vertical arrows of the left front square show how transition t is associated to a graph morphism via the g component of $\mathcal{M}$, and similarly for the back right square. Finally, the back left and front right sides of the top square are the components of the GTrS morphism associated to states s_1 and s_2, and the top square is required to commute.

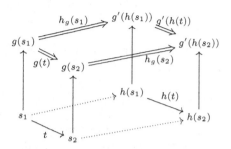

Definition 7 (unraveled graph transition system). *A GTrS $\mathcal{M} = \langle M, g \rangle$ is unraveled whenever M is a tree, for each $t \in T_M$ the morphism $g^T(t)$: $g^S(in_M(t)) \rightharpoonup g^S(out_M(t))$ is a partial inclusion, and item names are not re-used, i.e., for all $s', s'' \in S_M$, if $x \in X_{g^S(s')} \cap X_{g^S(s'')}$ there exists $s \in S_M$ such that*

$$x \in X_{g^S(s)} \wedge s \xrightarrow{*} s' \wedge s \xrightarrow{*} s'' \wedge g^T(s \xrightarrow{*} s')(x) = x \wedge g^T(s \xrightarrow{*} s'')(x) = x,$$

where $g^T(s \xrightarrow{*} s')$ is the composition of the partial morphisms associated with the transitions in $s \xrightarrow{*} s'$, which is uniquely determined since M is a tree.

An unraveling *of a GTrS $\mathcal{M} = \langle M, g \rangle$ is a pair $\langle \mathcal{M}', h \rangle$ where $\mathcal{M}'$ is an unraveled GTrS and $h = \langle h_M, h_g \rangle : \mathcal{M}' \to \mathcal{M}$ is a GTrS morphism, satisfying:*

1. *for each $s \in S_{M'}$, $h_g(s) : g'^S(s) \to g^S(h_M^S(s))$ is an isomorphism;*

2. *for each $s \in S_{M'}$ and transition $h_M^S(s) \xrightarrow{t'} s''$ in M, there is a transition $s \xrightarrow{t} s'$ in M' such that $h_M^T(t) = t'$ (and thus $h_M^S(s') = s''$).*

Proposition 1 (unraveling of a GTrS). *Any GTrS admits an unraveling.*

The conditions defining an unraveled GTrS $\mathcal{M}$ ensure that taking the union of all the graphs associated to the states in S_M, we obtain a well-defined graph. In fact, given any two states s and s' and an edge $e \in E_{g^S(s)} \cap E_{g^S(s')}$, the source, target and label of e coincide in $g^S(s)$ and $g^S(s')$. We shall denote the components of this "universe" graph as $\langle V_{\mathcal{M}}, E_{\mathcal{M}}, s_{\mathcal{M}}, t_{\mathcal{M}}, lab_{\mathcal{M}} \rangle$, where $V_{\mathcal{M}} = \bigcup_{s \in S_M} V_{g^S(s)}$, $E_{\mathcal{M}} = \bigcup_{s \in S_M} E_{g^S(s)}$, $s_{\mathcal{M}}(e) = s_{g^S(s)}(e)$ if $e \in g^S(s)$, and similarly for $t_{\mathcal{M}}$ and $lab_{\mathcal{M}}$.

3   A Temporal Graph Logic for Graph Transformation Systems

We now define syntax and semantics of our temporal graph logic, that extends the logic $\mu\mathcal{L}2$ of [3]. The logic is based both on the μ-calculus [6] and on second-order graph logic [8]. Let V_x, V_X, V_Z be sets of first- and second-order edge variables and propositional variables, respectively.

Definition 8 (syntax). *The logic $\mu\mathcal{G}2$ is given by the set of all formulae generated by:*

$$F ::= \eta(x) = \eta'(y) \mid x = y \mid l(x) = a \mid \neg F \mid F \vee F \mid \exists x.F \mid \exists X.F \mid$$
$$x \in X \mid Z \mid \Diamond F \mid \mu Z.F$$

where $\eta, \eta' \in \{s, t\}$ (standing for source and target), $x, y \in V_x$, $X \in V_X$, $a \in \Lambda$ and $Z \in V_Z$. Furthermore $\Diamond F$ is the (existential) next-step operator. The letter μ denotes the least fixed-point operator. As usual the formula $\mu Z.F$ can be formed only if all occurrences of Z in F are positive, i.e., they fall under an even number of negations. In the following we will use some (redundant) connectives like $\wedge$, $\forall$, $\Box$ and ν (greatest fixed-point), defined as usual. We denote by $\mu\mathcal{G}1$ its first-order fragment, where second-order edge variables and quantification are not allowed.

Definition 9 (semantics of $\mu\mathcal{G}2$). *Let $\mathcal{M} = \langle M, g \rangle$ be an unraveled GTrS. The semantics of temporal graph formulae is given by an evaluation function mapping closed formulae into subsets of S_M, i.e., the states that satisfy the formula. We shall define a mapping $[\![\cdot]\!]_\sigma^{\mathcal{M}} : \mu\mathcal{G}2 \to 2^{S_M}$, where σ is an environment, i.e., a tuple $\sigma = \langle \sigma_x, \sigma_X, \sigma_Z \rangle$ of mappings from first- and second-order edge variables into edges and edge sets, respectively, and from propositional variables into subsets of S_M. More precisely, $\sigma_x : V_x \to E_M$, $\sigma_X : V_X \to 2^{E_M}$ and $\sigma_Z : V_Z \to 2^{S_M}$, where E_M is the set of all edge names used in $\mathcal{M}$. When $\mathcal{M}$ is implicit, we simply write $[\![\cdot]\!]_\sigma$.*

$$[\![\eta(x) = \eta'(y)]\!]_\sigma = \lfloor\lfloor \eta_{\mathcal{M}}(\sigma_x(x)) = \eta'_{\mathcal{M}}(\sigma_x(y)) \rfloor\rfloor \qquad [\![x = y]\!]_\sigma = \lfloor\lfloor \sigma_x(x) = \sigma_x(y) \rfloor\rfloor$$
$$[\![l(x) = a]\!]_\sigma = \lfloor\lfloor lab_{\mathcal{M}}(\sigma_x(x)) = a \rfloor\rfloor \qquad [\![y \in Y]\!]_\sigma = \lfloor\lfloor \sigma_x(y) \in \sigma_X(Y) \rfloor\rfloor$$
$$[\![\neg F]\!]_\sigma = S_M \setminus [\![F]\!]_\sigma \qquad [\![F_1 \vee F_2]\!]_\sigma = [\![F_1]\!]_\sigma \cup [\![F_2]\!]_\sigma$$
$$[\![Z]\!]_\sigma = \sigma_Z(Z) \qquad [\![\mu Z.F]\!]_\sigma = \mathbf{lfp}(\lambda v.[\![F]\!]_{\sigma[v/Z]})$$

$$[\![\Diamond F]\!]_\sigma = \{s \in S_M \mid \exists s', t. s \xrightarrow{t} s' \wedge s' \in [\![F]\!]_\sigma\}$$
$$[\![\exists x.F]\!]_\sigma = \{s \in S_M \mid \exists e \in E_{g(s)} . s \in [\![F]\!]_{\sigma[e/x]}\}$$
$$[\![\exists X.F]\!]_\sigma = \{s \in S_M \mid \exists E \subseteq E_{g(s)} . s \in [\![F]\!]_{\sigma[E/X]}\}$$

where $\lfloor\lfloor \cdot \rfloor\rfloor$ maps true and false to S_M and $\emptyset$, respectively, $v \in 2^{S_M}$, and $\mathbf{lfp}(f)$ denotes the least fixed-point of the function f.

In particular, if F is a closed formula, we say that $\mathcal{M}$ satisfies F and write $\mathcal{M} \models F$, if $s_0 \in [\![F]\!]_\sigma$, where σ is the empty environment. Finally, we say that a GTS $\mathcal{R} = \langle G_0, R \rangle$ satisfies a closed formula F, written $\mathcal{R} \models F$, if the unraveling of GTrS($\mathcal{R}$) satisfies F.

The restriction to formulae where all occurrences of propositional variables are positive guarantees every possible function $\lambda v.[\![F]\!]_{\sigma[v/Z]}$ to be monotonic. Thus, by Knaster-Tarski theorem, fixed-points are well-defined.

Note that using unravelled GTrS is crucial for the definition of the semantics of the logic: items can be easily tracked since their identity is preserved and names are never reused. This allows to remember also the identities of deleted items, differently from what happens in the semantics given in [20,22].

Example 3. The following formula states that no M-labeled message edge is preserved by any transition: **M-consumed** $\equiv \neg \exists x.(l(x) = M \wedge \Diamond \exists y.\, x = y)$. The fact that this property holds in any reachable state is expressed by the formula: **always-M-consumed** $\equiv \nu Z.(\textbf{M-consumed} \wedge \Box Z)$. It is easy to see that **M-consumed** is satisfied by any state of the unraveled GTrS in Fig. 2, and thus $\mathcal{G} \models$ **always-M-consumed**, where $\mathcal{G}$ is the GTS of Fig. 1.

The formula **M-moves** $\equiv \neg \exists x.(l(x) = M \wedge \Diamond(\exists y.(l(y) = M \wedge t(y) = t(x) \wedge s(y) = s(x))))$ states that messages always move, i.e., there is no message edge such that in the next state there is another message edge with the same identity (i.e., source nodes coincide) attached to the same target node. And we can require this property to hold in every reachable state: **always-M-moves** $\equiv \nu Z.(\textbf{M-moves} \wedge \Box Z)$. Again, the GTS $\mathcal{G}$ satisfies this property. A GTS in which the message would at some point cross a "looping connection" or with more than one message would violate the formula.

4 Approximating GTSs with Petri Graphs

In the verification approach proposed in [1,3,4,5] *Petri graphs*, structures consisting of a Petri net and a graph component, have been introduced. They are used to represent finite approximations of the (usually infinite) unfolding of a GTS, on which to verify certain properties of the original system. Furthermore they provide a bridge to Petri net theory, allowing to reuse verification techniques developed for nets: a property expressed as a formula in the graph logic can be translated into an equivalent multiset formula to be verified on the net underlying the Petri graph. Here we shall concentrate on this latter aspect. We will not treat instead the construction of finite Petri graphs over-approximating GTSs, presented in [1,4] also for varying degrees of precision, recently enriched with a technique for counterexample-guided abstraction refinement [18], and for which the verification tool AUGUR (http://www.ti.inf.uni-due.de/research/augur_1/) has been developed.

Before introducing Petri graphs we need some definitions. Given a set A we will denote by $A^{\oplus}$ the free commutative monoid over A, whose elements will be called *multisets* over A. In the sequel we will sometimes identify $A^{\oplus}$ with the set of functions $m : A \to \mathbb{N}$ such that the set $\{a \in A \mid m(a) \neq 0\}$ is finite. E.g., in particular, $m(a)$ denotes the multiplicity of an element a in the multiset m. Sometimes a multiset will be identified with the underlying set, writing, e.g., $a \in m$ for $m(a) \neq 0$. Given a function $f : A \to B$, by $f^{\oplus} : A^{\oplus} \to B^{\oplus}$ we denote its monoidal extension, i.e., $f^{\oplus}(m)(b) = \sum_{f(a)=b} m(a)$ for every $b \in B$.

Definition 10 (Petri nets and Petri graphs). *A* (Place/Transition) *Petri net is a tuple* $N = \langle S_N, T_N, {}^{\bullet}(\ ), (\ )^{\bullet}, m_0 \rangle$, *where* S_N *is a set of* places, T_N *is a set of* transitions, ${}^{\bullet}(\ ), (\ )^{\bullet} : T_N \to S_N^{\oplus}$ *determine for each transition its* pre-set *and* post-set, *and* $m_0 \in S_N^{\oplus}$ *is the* initial marking. *A transition* t *is enabled at a*

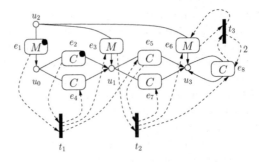

Fig. 3. A Petri graph

marking $m \in S_N^\oplus$ if $\bullet t \leq m$; in this case the transition can be fired at m, written $m[t\rangle m'$, and the resulting marking is $m' = m - \bullet t + t^\bullet$.[1]

A Petri graph P over a GTS $\mathcal{R} = \langle G_0, R \rangle$ is a tuple $\langle G, N, p \rangle$ where (1) G is a graph (sometimes called the template graph); (2) N is a Petri net such that (2.1) $S_N = E_G$, i.e., the set of places is the set of edges of G; (2.2) there is a graph morphism $\psi : G_0 \to G$; (2.3) the initial marking $m_0 \in E_G^\oplus$ properly corresponds to the start graph of $\mathcal{R}$, i.e., $m_0 = \psi^\oplus(E_{G_0})$; (3) $p : T_N \to R$ is a labeling function mapping each transition to a rule of $\mathcal{R}$, such that (3.1) for each transition $t \in T_N$, its pre- and post-sets $\bullet t$ and $t^\bullet$ properly correspond to the left- and right-hand side graphs of $p(t)$. A marking m is said to be reachable in P if there is a (possibly empty) sequence of firing $m_0[t_1\rangle m_1[t_2\rangle \ldots [t_n\rangle m$ in the underlying net N. The set of reachable markings of P is denoted by $\mathbf{M}_P$.

Example 4. Figure 3 depicts a Petri graph over the GTS $\mathcal{G}$ of our running example. It has been computed by the tool AUGUR as the covering of depth 1. As edges are places, the boxes representing edges can include tokens, which here represent the initial marking. Transitions are depicted as black rectangles. The pre-sets (resp. post-sets) of transitions are represented by dotted edges from edge places to transitions or vice versa. Note that all three transitions are instances of rule r, the only rule of $\mathcal{G}$, and that there is indeed a morphism from G_0 to the template graph, such that the image of this morphism (e_1, e_2) corresponds to the depicted (initial) marking.

A marking m of a Petri graph can be seen as an abstract representation of a graph. The intuition is that every token of an edge represents an instance of the corresponding template edge.

Definition 11 (graph generated by a marking). *Let $P = \langle G, N, p \rangle$ be a Petri graph and let $m \in E_G^\oplus$ be a marking of N. The* graph generated by m, *denoted by $graph_P(m)$, is the graph H defined as follows: $V_H = V_G$, $E_H = \{\langle e, i \rangle \mid e \in m \wedge 1 \leq i \leq m(e)\}$, $s_H(\langle e, i \rangle) = s_G(e)$, $t_H(\langle e, i \rangle) = t_G(e)$ and*

---

[1] Operations on markings are computed pointwise on the coefficients: e.g., $m_1 \leq m_2$ iff $m_1(s) \leq m_2(s)$ for all $s \in S_N$, and $(m_1 + m_2)(s) = m_1(s) + m_2(s)$.

$lab_H(\langle e, i \rangle) = lab_G(e)$. *In the following ψ_m denotes the obvious graph morphism from $graph_P(m)$ into G, which is the identity on nodes and which satisfies $\psi_{mE}(\langle e, i \rangle) = e \in E_G$.*

Example 5. The marking of the Petri graph depicted in Figure 3 generates the leftmost graph of Figure 4. The remaining graphs are generated by other reachable markings.

To each Petri graph P we can associate a GTrS, as shown in the next definition. As the reader would expect, each marking m is a state and the associated graph is $graph_P(m)$. However, each transition of P corresponds in general to a set of transitions in the GTrS, representing the different ways of preserving the edges. To see why this is necessary consider the Petri graph of our running example. Transition t_3 consumes a token from edge place e_8. Now assume that the marking m is such that $m(e_8) = 2$, i.e., there are two tokens in e_8. Such a marking is indeed reachable. In this case one has to consider two transitions in the transition system: one consuming edge $\langle e_8, 1 \rangle$ and the other consuming $\langle e_8, 2 \rangle$. The intuitive idea of having different ways of preserving edges is formalized by the notion of *significant preservations.*

Definition 12 (significant preservations). *Let $P = \langle G, N, p \rangle$ be a Petri graph, m, m' be markings and t a transition such that $m[t\rangle m'$. We denote by $SP(m, t)$ the set of significant preservations, which contains the possible different subsets of edges in $graph_P(m)$ which are not deleted by a firing of t, i.e., $SP(m, t) = \{ E_{graph_P(m)} - Y : Y \subseteq E_{graph_P(m)} \wedge \psi_m^{\oplus}(Y) = {}^{\bullet}t \}$.*

Definition 13 (GTrS by a Petri graph). *The GTrS generated by a Petri graph $P = \langle G, N, p \rangle$, denoted by $\mathrm{GTrS}(P)$, is $\langle M, g \rangle$ where*

- *S_M is the set of markings reachable in P, i.e., $S_M = \mathbf{M}_P$;*
- *$T_M = \{ \langle m, t, X, m' \rangle \mid m[t\rangle m' \text{ and } X \in SP(m, t) \}$;*
- *$in(\langle m, t, X, m' \rangle) = m$ and $out(\langle m, t, X, m' \rangle) = m'$ for $\langle m, t, X, m' \rangle \in T_M$;*
- *$s_0^M = m_0$;*
- *$g^S(m) = graph_P(m)$;*
- *$g^T(\langle m, t, X, m' \rangle) = f_{m,t,X}$, where $f_{m,t,X} : graph_P(m) \to graph_P(m')$ is any injective partial graph morphism which is the identity over nodes, and whose domain over edges is exactly X; for example, a concrete definition over edges can be $f_{m,t,X}^E(\langle e, i \rangle) = \langle e, k \rangle$ if $k = |\langle e, j \rangle \in X : j \leq i|$.*

Example 6. Figure 4 illustrates the GTrS associated to the Petri graph of our running example. For the sake of simplicity only the underlying graphs and partial graph morphisms are depicted. The leftmost state corresponds to the initial marking $m_0 = \{e_1, e_2\}$, while the next one corresponds to marking $m_1 = \{e_3, e_4, e_5\}$ reachable after firing t_1. In the third graph a looping C-edge is introduced, due to over-approximation. The most interesting transitions are t_4', t_4'', both leaving marking m_3. Both have the form $\langle m_3, t_3, X, m_4 \rangle$, where X can either be $\{\langle e_4, 1 \rangle, \langle e_7, 1 \rangle, \langle e_8, 1 \rangle\}$ or $\{\langle e_4, 1 \rangle, \langle e_7, 1 \rangle, \langle e_8, 2 \rangle\}$. These transitions represent different ways of consuming an instance of edge place e_8.

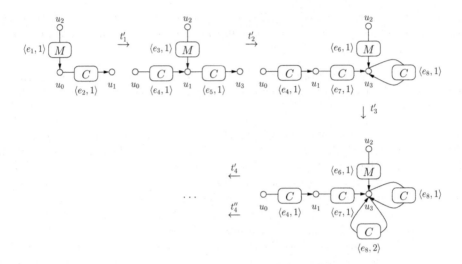

Fig. 4. A graph transition system of a Petri graph

Definition 14 (approximation). *Let $\mathcal{R} = \langle G_0, R \rangle$ be a GTS, let $P = \langle G, N, p \rangle$ be a Petri graph and let $\mathcal{M}_\mathcal{R}$ and $\mathcal{M}_P$ be the unravelings of the GTrSs generated by $\mathcal{R}$ and P, respectively. We say that P is an approximation of $\mathcal{R}$ if there is a GTrS morphism $\langle h_M, h_g \rangle : \mathcal{M}_\mathcal{R} \to \mathcal{M}_P$ such that for each state $s \in S_{\mathcal{M}_\mathcal{R}}$, the graph morphism $h_g(s)$ is edge-bijective.*

It is easy to see that this notion of approximation implies a simulation in Milner's sense: if the original system can make a transition t from a graph G to a graph H, then the approximation can simulate it with a transition from a graph G' to a graph H' via $h(t)$. Additionally, we require that there must be edge-bijective morphisms from G to G' and from H to H', a property which will be crucial for determining fragments of our logic that are preserved or reflected by the approximation.

Example 7. Consider our running example. It is easy to see that the Petri graph in Fig. 3 approximates the GTS in Fig. 1: there exist several morphisms with edge-bijective components from the GTrS of Fig. 2 to the unraveling of the GTrS of Fig. 4.

As already mentioned, given a GTS $\mathcal{R} = \langle G_0, R \rangle$, an algorithm proposed in [1,4] constructs a Petri graph associated to $\mathcal{R}$, called its *covering* and denoted by $\mathcal{C}(\mathcal{R})$. The covering is an approximation of $\mathcal{R}$ in the sense stated above.

Proposition 2 (covering approximates). *Let $\mathcal{R} = \langle G_0, R \rangle$ be a graph transformation system. Then the covering $\mathcal{C}(\mathcal{R}) = \langle G, N, p \rangle$ of $\mathcal{R}$ is an approximation of $\mathcal{R}$.*

5   Preservation and Reflection of Formulae

In this section we introduce a type system over graph formulae in $\mu\mathcal{G}2$, generalizing the one in [5], which characterizes subclasses of formulae preserved or reflected by approximations. More precisely, the type system may assign to a formula F the type "$\rightarrow$", meaning that F is preserved by approximations, or the type "$\leftarrow$", meaning that F is reflected by approximations. Note especially that since the approximation may merge nodes, formulae checking the identity of nodes are preserved, but not reflected.

Definition 15 (reflected/preserved formulae). *The typing rules are given by*

$$\eta(x) = \eta'(y): \rightarrow \quad x = y, \, l(x) = a, \, x \in X, Z: \leftrightarrow$$

$$\frac{F: d}{\neg F: d^{-1}} \quad \frac{F_1, F_2: d}{F_1 \vee F_2: d} \quad \frac{F: d}{\exists x.F: d} \quad \frac{F: d}{\exists X.F: d} \quad \frac{F: \rightarrow}{\Diamond F: \rightarrow} \quad \frac{F: \leftarrow}{\Box F: \leftarrow} \quad \frac{F: d}{\mu Z.F: d}$$

where it is intended that $\rightarrow^{-1} = \leftarrow$ and $\leftarrow^{-1} = \rightarrow$. Moreover $F: \leftrightarrow$ is a shortcut for $F: \rightarrow$ and $F: \leftarrow$, while $F_1, F_2: d$ stands for $F_1: d$ and $F_2: d$.

The type system can be shown to be correct in the following sense (see also [19]).

Proposition 3 (preservation and reflection). *Let $\mathcal{M}$ and $\mathcal{M}'$ be two unraveled GTrSs such that there is a morphism $\langle h_M, h_g \rangle : \mathcal{M} \to \mathcal{M}'$ having all h_g components edge-bijective. Then for each closed formula $F \in \mu\mathcal{G}2$ we have (i) if $F: \leftarrow$ then $\mathcal{M}' \models F$ implies $\mathcal{M} \models F$ and (ii) if $F: \rightarrow$ then $\mathcal{M} \models F$ implies $\mathcal{M}' \models F$.*

Not all formulae that are preserved respectively reflected are recognized by the above type system. A result of [5] shows that this incompleteness is a fundamental problem, due to the undecidability of reflection and preservation.

Example 8. Observe that **always-M-consumed** $: \leftarrow$ and hence approximations reflect this property. Indeed the unraveling of the GTrS of the Petri graph of our running example (see Fig. 4) satisfies the property, and so does the original GTrS. Also, formula **always-M-moves** is classified as reflecting: however, in this case the GTrS of Fig. 2 satisfies this property, but the unraveling of the GTrS of Figure 4 does not, by the presence of a connection loop e_8.

Since the covering provides an approximation of the original GTS, the theorem above applies. For a Petri graph P and a closed formula $F \in \mu\mathcal{G}2$, we shall write $P \models F$ if the unraveling of the graph transition system $GTrS(P)$ generated by P satisfies F.

Corollary 1 (covering preserves and approximates). *Let $\mathcal{R}$ be a GTS and let $F \in \mu\mathcal{G}2$ be a closed formula. Then we have (i) if $F: \leftarrow$ then $\mathcal{C}(\mathcal{R}) \models F$ implies $\mathcal{R} \models F$ and (ii) if $F: \rightarrow$ then $\mathcal{R} \models F$ implies $\mathcal{C}(\mathcal{R}) \models F$.*

6   From Temporal Graph Logics to Petri Net Logics

In this section we show how the first-order fragment of $\mu\mathcal{G}2$ can be encoded into a temporal logic for Petri nets, in a way that the Petri net underlying a Petri graph satisfies the encoding of a formula exactly when the Petri graph satisfies the original $\mu\mathcal{G}2$ formula.

Let V_Z be a set of propositional variables and N_P, N_T sets from which place and transition names are drawn, respectively.

Definition 16 (Petri net logic syntax). *The syntax of the Petri net logic* $\mathcal{P}$ *is given by the following grammar, where* $p \in N_P$, $t \in N_T$, $c \in \mathbb{N}$ *and* $Z \in V_Z$:

$$\phi ::= \#p \leq c \mid \phi \vee \phi \mid \neg\phi \mid Z \mid \mu Z.\phi \mid \langle t \rangle \phi.$$

The semantics is mostly standard and given by a mapping $[\![\cdot]\!]_\sigma^P : \mathcal{P} \to 2^{\mathbf{M}_P}$ mapping formulae into sets of markings, where $\sigma : free_Z \to 2^{\mathbf{M}_P}$ is an environment mapping propositional variables into sets of markings. We sometimes use a satisfaction relation $\models_\sigma \subseteq \mathbf{M}_P \times \mathcal{P}$, where $m \models_\sigma \phi$ whenever $m \in [\![\phi]\!]_\sigma^P$.

As an example, $\#e \leq c$ is satisfied by markings m, where $m(e) \leq c$, i.e., markings where the number of tokens in place e is less than or equal to c. Next-time modalities are enriched with transition labels with the following meaning: $\langle t \rangle \phi$ is satisfied by markings from which transition t can be fired leading to a marking that satisfies ϕ.

The encoding $[\![\cdot]\!]$, which maps formulae of $\mu\mathcal{G}1$ into formulae of the logic $\mathcal{P}$, is based on the following observation: Every graph $graph_P(m)$ for some marking m of P can be generated from the finite template graph G in the following way: some edges of G might be removed and some edges be multiplied, generating several parallel instances of the same template edge. Whenever a formula has two free variables x, y and $graph_P(m)$ has n parallel instances $e_1, \ldots, e_n$ of the same edge, it is not necessary to associate x and y to edges in all possible ways, but it is sufficient to remember whether x and y are mapped or not mapped to the same edge. Hence, in the encoding of a formula F, we keep track of the following information: a partition Q on the free variables $free(F)$, telling which variables are mapped to the same edge, and a mapping ρ from $free(F)$ to the edges of G, with $\rho(x) = e$ meaning that x will be instantiated with an instance of the template edge e. When encoding an existential quantifier $\exists x$, we form a disjunction over all the possibilities we have in choosing such an x: either x is instantiated with the same edge as another free variable y, and thus x and y are in the same class of the partition Q; or x is mapped to a new instance of an edge in G, and thus a new set $\{x\}$ is added to Q, adding a suitable predicate which ensures that enough edges are available.

This is enough for the logic of [5], where interleaving of temporal operators and edge quantifiers is not allowed. Here we have to consider the case in which temporal modalities are nested into edge quantification. The main problem is that an edge where some variables have been mapped can be removed by a transition and thus, when encoding quantification, one must be careful in avoiding to instantiate a variable with the class of a removed edge. This is faced by recording in a set R the classes corresponding to *removed* edges.

Before we define the encoding we need some definitions. An equivalence relation Q over a set A will be represented as a partition $Q \subseteq 2^A$, where every element represents an equivalence class. We will write xQy whenever x, y are in the same equivalence class $k \in Q$. Furthermore we assume that each equivalence Q is associated with a function $rep : Q \to A$ which assigns a representative to every equivalence class. The encoding given below is independent of any specific choice of representatives. Given a function $f : A \to B$ such that $f(a) = f(a')$ for all $a, a' \in A$ with aQa', we shall often write $f(k)$ for $f(rep(k))$; furthermore, for any $b \in B$ we define $n_{Q,f}(b) = |\{k \in Q \mid f(k) = b\}|$, i.e., the number of classes in the partition Q that are mapped to b.

We next define the encoding, concentrating first on the fragment without fixed-point operators.

Definition 17 (encoding for the fixed-point free first-order logic). *Let $P = \langle G, N, p \rangle$ be a Petri graph, F be a fixed-point-free $\mu\mathcal{G}1$ formula, $\rho : free(F) \to E_G$ and $Q \subseteq 2^{free(F)}$ be an equivalence relation, $R \subseteq Q$ and xQy implies $\rho(x) = \rho(y)$ for all $x, y \in free(F)$. The encoding $[\cdot] : \mu\mathcal{G}1 \to \mathcal{P}$ is defined as follows:*

$$[\neg F, \rho, Q, R] = \neg[F, \rho, Q, R]$$
$$[F_1 \vee F_2, \rho, Q, R] = [F_1, \rho, Q, R] \vee [F_2, \rho, Q, R]$$
$$[x = y, \rho, Q, R] = \begin{cases} true & if\ xQy \\ false & otherwise \end{cases}$$
$$[l(x) = a, \rho, Q, R] = \begin{cases} true & if\ lab_G(\rho(x)) = a \\ false & otherwise \end{cases}$$
$$[s(x) = s(y), \rho, Q, R] = \begin{cases} true & if\ s_G(\rho(x)) = s_G(\rho(y)) \\ false & otherwise \end{cases}$$
$$analogously\ for\ t(x) = t(y)\ and\ s(x) = t(y)$$
$$[\exists x.F, \rho, Q, R] = \bigvee_{k \in Q \backslash R}[F, \rho \cup \{x \mapsto \rho(k)\}, Q \setminus \{k\} \cup \{k \cup \{x\}\}, R] \vee$$
$$\vee \bigvee_{e \in E_G}([F, \rho \cup \{x \mapsto e\}, Q \cup \{\{x\}\}, R]$$
$$\wedge (\#e \geq n_{Q \backslash R, \rho}(e)+1))$$
$$[\Diamond F, \rho, Q, R] = \bigvee_{t \in T_N} \bigvee_{R' \in S_{R,t}} (\bigwedge_{e \in \ {}^\bullet t}(\#e \geq n_{Q \backslash (R \cup R'), \rho}(e)+ {}^\bullet t(e)) \wedge$$
$$\wedge \langle t \rangle [F, \rho, Q, R \cup R'])$$
$$[Z, \rho, Q, R] = Z$$

where $S_{R,t}$ abbreviates $\{R' \in 2^{(Q \backslash R)} \mid (\rho \circ rep)^\oplus(R') \leq\ {}^\bullet t\}$.

If F is closed, we define $[F]$ to be $[F, \emptyset, \emptyset, \emptyset]$. The main novelty with respect to [5] is the encoding of formulae $\Diamond F$ involving the next-time operator. In order to see if there is a transition after which F holds we examine the possible transitions t of the Petri graph, and hence the disjunction amongst all $t \in T_N$ arises. Concerning the removed edges, after the firing of a transition t several cases may apply since edges corresponding to places in the pre-set of t may be preserved or consumed, depending on the number of tokens in such places.

All the cases that have to be considered are defined by $S_{R,t}$. In words, $S_{R,t}$ contains sets of equivalence classes from $Q \setminus R$ that correspond to places in the preset of t, not exceeding the number of tokens removed by t from each place, i.e., if a transition consumes n tokens from a place e we shall not consider the case in which more than n equivalence classes mapped to e are consumed.

When considering the consumption of one of such set R' of equivalence classes, we have to ensure that equivalence classes not included in R' can actually be preserved. This can only happen if there are enough tokens ($\bigwedge_{e \in \bullet t} \#e \geq n_{Q \setminus (R \cup R'), \rho}(e) + {}^\bullet t(e)$), i.e., if in each place e in the pre-set of t the number of tokens is greater or equal to the number of tokens removed by t from e plus the number of equivalence classes mapped to e that will be preserved.

Example 9. To clarify this point consider an example in which we are treating a transition t whose pre-set is $\{e, e', e'\}$ and we have that Q is $\{k_1, k_2, k_3\}$, R is empty and ρ maps k_1, k_2 to e and k_3 to e'. In this case we have that $S_{R,t}$ is $\{\emptyset, \{k_1\}, \{k_2\}, \{k_1, k_3\}, \{k_2, k_3\}\}$. Note that $\{k_1, k_2, k_3\}$ does not belong to $S_{R,t}$ because the in-degree of e for t is just 1. Now let us consider the requirements on the places for some of the elements of $S_{R,t}$. For instance, for $\{k_1\}$ we need $\#e \geq 2 \wedge \#e' \geq 3$ in order to be able to preserve k_2 and k_3, while for $\{k_1, k_3\}$ we need $\#e \geq 2$ to preserve k_2.

After fixing an R' and setting the requirement on the number of tokens in the pre-set of t we have only to state that formula F holds under the new configuration, i.e., where R' is added to the set of removed equivalence classes.

Example 10. For the Petri graph in Fig. 3 formula [**M-consumed**] has the following form:

$$\neg \bigvee_{(i,j) \in \mathcal{L}} \left((\#e_i \geq 2 \wedge \#e_j \geq 1) \vee \left(\#e_i \geq 1 \wedge \bigvee_{(k,\ell) \in \mathcal{L} - \{(i,j)\}} (\#e_k \geq 1 \wedge \#e_\ell \geq 1) \right) \right)$$

where $\mathcal{L} = \{(1,2), (3,5), (6,8)\}$ (pairs of indices of edges which form the pre-set of a transition). Intuitively, the two disjuncts above encode the fact that an M-edge can be preserved if there is an enabled transition with more than one token in the M-edge in its pre-set (left disjunct), or there is a token in an M-edge and a transition which does not consume this token is enabled (right disjunct).

In the worst case there can be an exponential blowup in the size of the encoded formula. But at the same time, the resulting formula can often be greatly simplified, even on-the-fly.

The encoding for general formulae of $\mu\mathcal{G}1$, possibly including fixed-point operators requires additional effort. Suppose we want to express that there is an edge x and a computation that never consumes x, i.e., $F \equiv \exists x.\nu Z.\exists y.(x = y \wedge \Diamond Z)$.

Now, if we try to encode F we encounter a problem: sometimes Z should be evaluated in a context where the equivalence class of x is preserved and sometimes in one where x is consumed.

In order to solve this, we exploit a property of fixed-points, namely that unfolding an occurrence of the variable of a fixed-point formula results in an equivalent formula. More formally, $\mu Z.F$ is equivalent to $\mu Z.F'$, where F' is the same as F except that some free occurrences of Z are substituted by $\mu Z'.F[Z'/Z]$, where Z' is a fresh variable. As we shall see, unfolded fixed-points will be evaluated in different contexts and a syntactic restriction will guarantee termination of the encoding.

This is handled in the encoding of the next-step operator where in the case of partition-consuming transitions we use the unfolding of F, denoted by $uf(F, R)$. The unfolding is formally defined as follows. Let $fp(Z)$ denote the fixed-point formula corresponding to a propositional variable Z and let $\{Z_1^F, .., Z_n^F\}$ denote the set of propositional variables appearing free in F. Then $uf(F, R)$ is defined as $F\{fp(Z_1^F)/Z_1^F, .., fp(Z_n^F)/Z_n^F\}$, i.e., each variable is substituted by the corresponding fixed-point formula, if R is not empty, otherwise $uf(F, R)$ is just F. The idea is that if no equivalence class is consumed the unfolding is not necessary. Of course, every propositional and edge variable must be renamed in the unfolding.

Formally, the encoding in presence of fixed-point operators is defined as follows.

Definition 18 (encoding for first-order logic). *Let $P = \langle G, N, p \rangle$ be a Petri graph, $F \in \mu\mathcal{G}1$, ρ, Q, R as in Definition 17. The encoding $[\cdot] : \mu\mathcal{G}1 \to \mathcal{P}$ is defined as in Definition 17, but for the clause of the next-step operator which becomes (a) below and the new clause for fixed-point operators (b):*

$$[\lozenge F, \rho, Q, R] = \bigvee_{t \in T_N} \bigvee_{R' \in S_{R,t}} (\bigwedge_{e \in \, {}^\bullet t} (\#e \geq n_{Q \setminus (R \cup R'), \rho}(e) + {}^\bullet t(e)) \wedge \qquad (a)$$
$$\wedge \langle t \rangle [uf(F, R'), \rho, Q, R \cup R'])$$

$$[\mu Z.\phi, \rho, Q, R] = \mu Z.[\phi, \rho, Q, R] \qquad\qquad\qquad\qquad (b)$$

In order to guarantee termination of the encoding, we have to forbid formulae in which propositional variables appear free under the scope of an edge quantifier. To see why this is necessary, one can apply the encoding to formula $\nu Z.\exists x.\lozenge(Z \wedge \exists y.y = x)$, expressing that there is an infinite computation where in every state there is at least one edge that is preserved in the next state. With the restriction mentioned the encoding is guaranteed to terminate, since in this case the set Q will not increase, hence set $S_{R,t}$ will decrease and at some point the chosen R' must be empty. It is an open question whether this syntactic restriction involves a loss of expressive power.

Proposition 4 (finite encoding). *Let (G, N, p) be a Petri graph, $F \in \mu\mathcal{G}1$ such that no propositional variable appears free under the scope of an edge quantifier. Then $[F, \emptyset, \emptyset, \emptyset]$ is finite.*

Example 11. We have that [**always-M-consumed**] equals

$$
\nu Z. \left([\textbf{M-consumed}] \wedge \bigwedge_{(i,j,k)\in\mathcal{M}} (\#e_i \geq 1 \wedge \#e_j \geq 1 \Rightarrow [t_k]Z) \right),
$$

where $\mathcal{M} = \{(1,2,1),(3,5,2),(6,8,3)\}$ (indices of pre-sets together with transition indices) and $[t]\phi = \neg\langle t\rangle\neg\phi$. That is, in order to ensure that it is impossible for a message to be preserved in any reachable state we require that **M-consumed** holds for any reachable marking.

This formula is satisfied by the Petri net underlying the Petri graph in Fig. 3 and it can even be verified using standard techniques for coverability checking.

Finally, we state correctness of the encoding. This result, together with Corollary 1 allows to check that a formula F in $\mu\mathcal{G}1$, typed as reflected, holds for a GTS $\mathcal{R}$ by checking that its encoding $[F]$ holds in the Petri net underlying any covering of $\mathcal{R}$.

Proposition 5 (correct encoding). *Let $P = \langle G, N, p\rangle$ be a Petri graph and let F be a closed formula in $\mu\mathcal{G}1$. Then $m_0 \models_\emptyset [F, \emptyset, \emptyset, \emptyset]$ iff $P \models F$.*

7   Conclusion

We have enriched an existing approach for the verification of behavioral properties of GTSs via approximation [1,4,5]. The original approach approximates a GTS by a Petri graph and reduces temporal graph formulae to existing logics for Petri nets. The original logic proposed in [5] does not allow to interleave temporal modalities and edge quantifiers and is thus not able to express properties like *an edge is never removed*. We have proposed a solution to this, by using a logic that interleaves temporal and structural aspects of a GTS and extending the encoding into Petri net logics.

Our work is not the first one that proposes a non-propositional temporal logic for graph transformation systems. The most relevant approach, in this respect, is [20], where a second-order LTL logic is proposed, which is interpreted on "graph transition systems" (these however are defined in a slightly different way). Our approach is more general in the sense that also consider systems with a possibly infinite state space and approximations of these systems, whereas [20] considers finite-state systems; furthermore the temporal aspect of our logic subsumes LTL. However, a precise comparison of the two approaches is not easy, because the graph-related aspect of the logics are different ([20] considers a logic to express path properties as regular expressions, while ours is based on a fragment of the monadic second-order logic of graphs [8]), and graph items which are deleted are handled differently in the two approaches.

Another first order temporal logic—called *evolution logic*—is proposed in [26], in a framework based on abstract interpretation for the verification of Java programs featuring dynamic allocation and deallocation of objects and threads.

Evolution logic is a first order version of LTL, enriched with transitive closure, and we think it suitable to express complex properties of graph transformation systems as well: a deep comparison in this respect with our logic $\mu\mathcal{G}2$ is planned as future work.

In [22] it has been shown how the verification problem for CTL with additional quantification over "items" can be reduced to the verification of standard CTL. This is not directly applicable to our setting, since we consider infinite-state systems, but the connection to [22] deserves further study. Another related work is [9], which is concerned with the approximation of special kinds of graphs and the verification of a similar logic for verifying pointer structures on a heap.

In future work we plan to study the decidability of fragments of our logic. First, we can profit from decidability results on similar logics like the guarded monadic fragments of first-order temporal logics [16] and similar approaches for the modal μ-calculus with first-order predicates [13]. Note that from a practical point of view we can focus on the target Petri net logic $\mathcal{P}$. Although the full logic is undecidable, there are some clearly identifiable decidable fragments [12,17]. In the linear-time case for instance, it is decidable to show whether there exists a run satisfying a formula containing only "eventually" operators, but mixing of "eventually" and "generally" operators in general leads to an undecidable logic.

Additionally, further approximation on the transition system generated by the Petri net can be used in order to model-check formulas on infinite-state Petri nets (see, e.g., [24]). We plan to enhance our approach by extending the encoding to the full logic including second-order quantification, and considering more general graph transformation systems, allowing non-discrete interfaces [2].

References

1. P. Baldan, A. Corradini, and B. König. A static analysis technique for graph transformation systems. In *CONCUR'01*, volume 2154 of *Lecture Notes in Computer Science*. Springer, 2001.
2. P. Baldan, A. Corradini, and B. König. Verifying finite-state graph grammars: An unfolding-based approach. In *Proc. of CONCUR'04*, volume 3170 of *Lecture Notes in Computer Science*, pages 83–98. Springer, 2004.
3. P. Baldan, A. Corradini, B. König, and B. König. Verifying a behavioural logic for graph transformation systems. In *Proc. of COMETA'03*, volume 104 of *ENTCS*, pages 5–24. Elsevier, 2004.
4. P. Baldan and B. König. Approximating the behaviour of graph transformation systems. In *Proc. of ICGT'02*, volume 2505 of *Lecture Notes in Computer Science*, pages 14–29. Springer, 2002.
5. P. Baldan, B. König, and B. König. A logic for analyzing abstractions of graph transformation systems. In *SAS'03*, volume 2694 of *Lecture Notes in Computer Science*, pages 255–272. Springer, 2003.
6. J. Bradfield and C. Stirling. Modal logics and mu-calculi: an introduction. In J. Bergstra, A. Ponse, and S. Smolka, editors, *Handbook of Process Algebra*. Elsevier, 2001.
7. A. Corradini, U. Montanari, F. Rossi, H. Ehrig, R. Heckel, and M. Löwe. *Algebraic Approaches to Graph Transformation I: Basic Concepts and Double Pushout Approach*, chapter 3. Volume 1 of Rozenberg [23], 1997.

8. B. Courcelle. *The expression of graph properties and graph transformations in monadic second-order logic*, chapter 5, pages 313–400. Volume 1 of Rozenberg [23], 1997.
9. D. Distefano, J.-P. Katoen, and A. Rensink. Who is pointing when to whom? In K. Lodaya and M. Mahajan, editors, *Proc. of FSTTCS'04*, volume 3328 of *Lecture Notes in Computer Science*, pages 250–262. Springer, 2004.
10. F. Dotti, L. Foss, L. Ribeiro, and O. Marchi Santos. Verification of distributed object-based systems. In *Proc. of FMOODS '03*, volume 2884 of *Lecture Notes in Computer Science*, pages 261–275. Springer, 2003.
11. H. Ehrig, R. Heckel, M. Korff, M. Löwe, L. Ribeiro, A. Wagner, and A. Corradini. *Algebraic Approaches to Graph Transformation II: Single Pushout Approach and Comparison with Double Pushout Approach*, chapter 4. Volume 1 of Rozenberg [23], 1997.
12. J. Esparza and M. Nielsen. Decidability issues for Petri nets - a survey. *Journal of Information Processing and Cybernetic*, 30(3):143–160, 1994.
13. E. Franconi and D. Toman. Fixpoint extensions of temporal description logics. In *Proc. of the International Workshop on Description Logics (DL2003)*, volume 81 of *CEUR Workshop Proceedings*, 2003.
14. F. Gadducci, R. Heckel, and M. Koch. A fully abstract model for graph intepreted temporal logic. In *Proc. of TAGT'98*, volume 1764 of *Lecture Notes in Computer Science*, pages 310–322. Springer, 1998.
15. M. Große-Rhode. Algebra transformation systems as a unifying framework. *Electronic Notes in Theoretical Computer Science*, 51, 2001.
16. I. Hodkinson, F. Wolter, and M. Zakharyaschev. Monadic fragments of first-order temporal logics: 2000-2001 a.d. In *Proc. of LPAR'01*, pages 1–23. Springer, 2001.
17. R. R. Howell, L. E. Rosier, and H.-C. Yen. A taxonomy of fairness and temporal logic problems for Petri nets. *Theoretical Computer Science*, 82:341–372, 1991.
18. B. König and V. Kozioura. Counterexample-guided abstraction refinement for the analysis of graph transformation systems. In *Proc. of TACAS '06*, volume 3920 of *Lecture Notes in Computer Science*, pages 197–211. Springer, 2006.
19. C. Loiseaux, S. Grafa, J. Sifakis, A. Bouajjani, and S. Bensalem. Property preserving abstractions for the verification of concurrent systems. *Formal Methods in System Design*, 6:1–35, 1995.
20. A. Rensink. Towards model checking graph grammars. In *Proc. of the 3rd Workshop on Automated Verification of Critical Systems*, Technical Report DSSE–TR–2003–2, pages 150–160. University of Southampton, 2003.
21. A. Rensink. Canonical graph shapes. In *Proc. of ESOP '04*, volume 2986 of *Lecture Notes in Computer Science*, pages 401–415. Springer, 2004.
22. A. Rensink. Model checking quantified computation tree logic. In *Proc. of CONCUR '06*, volume 4137 of *Lecture Notes in Computer Science*, pages 110–125. Springer, 2006.
23. G. Rozenberg, editor. *Handbook of Graph Grammars and Computing by Graph Transformation: Foundations*, volume 1. World Scientific, 1997.
24. K. Schmidt. Model-checking with coverability graphs. *Formal Methods in System Design*, 15(3), 1999.
25. D. Varró. Automated formal verification of visual modeling languages by model checking. *Software and System Modeling*, 3(2):85–113, 2004.
26. E. Yahav, T. Reps, M. Sagiv, and R. Wilhelm. Verifying temporal heap properties specified via evolution logic. In *Proc. of ESOP '03*, volume 2618 of *Lecture Notes in Computer Science*, pages 204–222. Springer, 2003.

On the Algebraization of Many-Sorted Logics*

Carlos Caleiro and Ricardo Gonçalves

CLC and SQIG-IT
Department of Mathematics, IST, TU Lisbon, Portugal

Abstract. The theory of abstract algebraic logic aims at drawing a strong bridge between logic and universal algebra, namely by generalizing the well known connection between classical propositional logic and Boolean algebras. Despite of its successfulness, the current scope of application of the theory is rather limited. Namely, logics with a many-sorted language simply fall out from its scope. Herein, we propose a way to extend the existing theory in order to deal also with many-sorted logics, by capitalizing on the theory of many-sorted equational logic. Besides showing that a number of relevant concepts and results extend to this generalized setting, we also analyze in detail the examples of first-order logic and the paraconsistent logic $\mathcal{C}_1$ of da Costa.

1 Introduction

The general theory of *abstract algebraic logic* (AAL from now on) was first introduced in [1]. It aims at providing a strong bridge between logic and universal algebra, namely by generalizing the so-called *Lindenbaum-Tarski method*, which led to the well known connection between classical propositional logic and Boolean algebras. Within AAL, one explores the relationship between a given logic and a suitable algebraic theory, in a way that enables one to use algebraic tools to study the metalogical properties of the logic being algebraized, namely with respect to axiomatizability, definability, the deduction theorem, or interpolation [2]. Nevertheless, AAL has only been developed (as happened, until recently, also with much of the research in universal algebra) for the single-sorted case. This means that the theory applies essentially only to propositional-based logics, and that logics over many-sorted languages simply fall out of its scope. It goes without saying that rich logics, with many-sorted languages, are essential to specify and reason about complex systems, as also argued and justified by the theory of combined logics [3,4].

Herein, we propose a way to extend the scope of applicability of AAL by generalizing to the many-sorted case several of the key concepts and results of the current theory, including several alternative *characterization results*, namely

---

* This work was partially supported by FCT and FEDER through POCI, namely via the QuantLog POCI/MAT/55796/2004 Project of CLC and the recent KLog initiative of SQIG-IT. The second author was also supported by FCT under the PhD grant SFRH/BD/18345/2004/SV7T and a PGEI research grant from Fundação Calouste Gulbenkian.

J.L. Fiadeiro and P.-Y. Schobbens (Eds.): WADT 2006, LNCS 4409, pp. 21–36, 2007.
© Springer-Verlag Berlin Heidelberg 2007

those involving the *Leibniz operator* and *maps of logics*. The generalization we propose assumes that the language of a logic is built from a many-sorted signature, with a distinguished sort for formulas. The algebraic counterpart of such logics will then be obtained via a strong representation over a suitable many-sorted algebraic theory, thus extending the notion of single-sorted algebraization of current AAL. Terms of other sorts may exist, though, but they do not correspond to formulas. Of course, in a logic, one reasons only about formulas, and only indirectly about terms of other sorts. Hence, we consider that the sort of formulas is the only visible sort, and we will also aim at a direct application of the theory of *hidden algebra*, as developed for instance in [5], to explore possible behavioral characterizations of the algebraic counterpart of a given many-sorted algebraizable logic.

We explore the new concepts by analyzing the example of first-order logic in a many-sorted context, and comparing with its previous unsorted study [1,6]. We will also see how to apply our many-sorted approach in order to provide a new algebraic perspective to certain logics which are not algebraizable in current AAL. Namely, we will establish the many-sorted algebraization of the paraconsistent logic C_1 of da Costa [7], whose single-sorted non-algebraizability is well known [8,9].

The paper is organized as follows. In section 2 we will introduce a number of necessary preliminary notions and notations. In section 3 we will introduce the essential concepts and results of current AAL, and present some of its limitations by means of examples. Then, in section 4, we will present our generalized notion of many-sorted algebraizable logic and a detailed analysis of the examples in the generalized setting. We will also show that some relevant concepts and results of AAL smoothly extend to the many-sorted setting. Finally, section 5 draws some conclusions, and points to several topics of future research.

2   Preliminaries

In this section we introduce the preliminary notions and notations that we will need in the rest of the paper, namely concerning logic and algebra.

2.1   Logics and Maps

We will adopt the Tarskian notion of logic. A logic is a pair $\mathcal{L} = \langle L, \vdash \rangle$, where L is a set of formulas and $\vdash\, \subseteq 2^L \times L$ is a consequence relation satisfying the following conditions, for every $\Gamma \cup \Phi \cup \{\varphi\} \subseteq L$:

Reflexivity: if $\varphi \in \Gamma$ then $\Gamma \vdash \varphi$;
Cut: if $\Gamma \vdash \varphi$ for all $\varphi \in \Phi$, and $\Phi \vdash \psi$ then $\Gamma \vdash \psi$;
Weakening: if $\Gamma \vdash \varphi$ and $\Gamma \subseteq \Phi$ then $\Phi \vdash \varphi$.

We will consider only these three conditions. However, Tarski also considered a *finitariness* condition (see [10]):

Finitariness: if $\Gamma \vdash \varphi$ then $\Gamma' \vdash \varphi$ for some finite $\Gamma' \subseteq \Gamma$.

In the sequel if $\Gamma, \Phi \subseteq L$, we shall write $\Gamma \vdash \Phi$ whenever $\Gamma \vdash \varphi$ for all $\varphi \in \Phi$. We say that φ and ψ are *interderivable*, which is denoted by $\varphi \dashv\vdash \psi$, if $\varphi \vdash \psi$ and $\psi \vdash \varphi$. In the same way, given $\Gamma, \Phi \subseteq L$ we say that Γ and Φ are *interderivable*, if $\Gamma \vdash \Phi$ and $\Phi \vdash \Gamma$. The *theorems* of $\mathcal{L}$ are the formulas φ such that $\emptyset \vdash \varphi$. A *theory* of $\mathcal{L}$, or briefly a $\mathcal{L}$-*theory*, is a set Γ of formulas such that if $\Gamma \vdash \varphi$ then $\varphi \in \Gamma$. Given a set Γ, we can consider the set $\Gamma^\vdash$, the smallest theory containing Γ. The set of all theories of $\mathcal{L}$ is denoted by $Th_\mathcal{L}$. It is easy to see that $\langle Th_\mathcal{L}, \subseteq \rangle$ forms a complete partial order.

We will need to use a rather strong notion of map of logics. Let $\mathcal{L} = \langle L, \vdash \rangle$ and $\mathcal{L}' = \langle L', \vdash' \rangle$ be two logics. A map θ from $\mathcal{L}$ to $\mathcal{L}'$ is a function $\theta : L \to 2^{L'}$ such that, if $\Gamma \vdash \varphi$ then $(\bigcup_{\gamma \in \Gamma} \theta(\gamma)) \vdash' \theta(\varphi)$. The map θ is said to be *conservative* when $\Gamma \vdash \varphi$ iff $(\bigcup_{\gamma \in \Gamma} \theta(\gamma)) \vdash' \theta(\varphi)$. A strong representation of $\mathcal{L}$ in $\mathcal{L}'$ is a pair (θ, τ) of conservative maps $\theta : \mathcal{L} \to \mathcal{L}'$ and $\tau : \mathcal{L}' \to \mathcal{L}$ such that:

i) For all $\varphi \in L$ we have that $\varphi \dashv\vdash \tau[\theta(\varphi)]$;
ii) For all $\varphi' \in L'$ we have that $\varphi' \dashv\vdash' \theta[\tau(\varphi')]$.

Note that the cases where $\theta(\varphi)$ is a singleton set for every $\varphi \in L$, or is a finite set for every $\varphi \in L$, are usual particular cases of the above definition of map. For the sake of notation we will use $\theta[\Gamma] = \bigcup_{\gamma \in \Gamma} \theta(\gamma)$. Hence, a map θ is such that if $\Gamma \vdash \varphi$ then $\theta[\Gamma] \vdash \theta(\varphi)$. Analogously, it is conservative when $\Gamma \vdash \varphi$ iff $\theta[\Gamma] \vdash \theta(\varphi)$. Note also that, because of the symmetry of the definition of strong representation, we also have a strong representation (τ, θ) of $\mathcal{L}'$ in $\mathcal{L}$. The existence of a strong representation of $\mathcal{L}$ in $\mathcal{L}'$ intuitively means that the consequence relation of $\mathcal{L}$ can be represented in $\mathcal{L}'$, and vice-versa, such that they are, in some precise sense, inverse of each other. Actually, θ and τ induce an isomorphism of the complete partial orders of theories of $\mathcal{L}$ and $\mathcal{L}'$. It is not difficult to see that, if we assume the conservativeness of θ and consider any function $\tau : L' \to 2^L$ that satisfies **ii)**, then we can conclude that τ is in fact a conservative map from $\mathcal{L}'$ to $\mathcal{L}$ that also satisfies **i)**.

2.2   Algebra

Recall that a many-sorted signature is a pair $\Sigma = \langle S, O \rangle$ where S is a set (of *sorts*) and $O = \{O_{ws}\}_{w \in S^*, s \in S}$ is an indexed family of sets. For simplicity, we write $f : s_1 \ldots s_n \to s$ for an element f of $O_{s_1 \ldots s_n s}$. As usual, we denote by $T_\Sigma(X)$ the S-indexed family of carrier sets of the free Σ-algebra $\mathbf{T}_\Sigma(\mathbf{X})$ with generators taken from a sorted family $X = \{X_s\}_{s \in S}$ of variable sets. Often, we will need to write terms over a given finite set of variables $t \in T_\Sigma(x_1 : s_1, \ldots, x_n : s_n)$. For simplicity, we will denote such a term by $t(x_1 : s_1, \ldots, x_n : s_n)$. Moreover, if T is a set whose elements are all terms of this form, we will denote this fact by writing $T(x_1 : s_1, \ldots, x_n : s_n)$. Fixed X, we will use $t_1 \approx t_2$ to represent an equation $\langle t_1, t_2 \rangle$ between Σ-terms t_1 and t_2 of the same sort[1]. If t_1 and t_2 are

---

[1] We use the symbol $\approx$ to avoid confusion with the usual symbol $=$ for (metalevel) equality, as used in the definitions.

terms of sort s, we will dub $t_1 \approx t_2$ an s-equation. The set of all Σ-equations will be written as Eq_Σ. Moreover, we will denote conditional equations by $t_1 \approx u_1 \wedge \cdots \wedge t_n \approx u_n \rightarrow t \approx u$. A set Θ whose elements are all equations over terms of the form $t(x_1 : s_1, \ldots, x_n : s_n)$, will also be dubbed $\Theta(x_1 : s_1, \ldots, x_n : s_n)$. A substitution $\sigma = \{\sigma_s : X_s \rightarrow T_\Sigma(X)_s\}_{s \in S}$ is an indexed family of functions. As usual, $\sigma(t)$ denotes the term obtained by uniformly applying σ to each variable in t. Given $t(x_1 : s_1, \ldots, x_n : s_n)$ and terms $t_1 \in T_\Sigma(X)_{s_1}, \ldots, t_n \in T_\Sigma(X)_{s_n}$, we will write $t(t_1, \ldots, t_n)$ to denote the term $\sigma(t)$ where σ is a substitution such that $\sigma_{s_1}(x_1) = t_1, \ldots, \sigma_{s_n}(x_n) = t_n$. Extending everything to sets of terms, given $T(x_1 : s_1, \ldots, x_n : s_n)$ and $U \in T_\Sigma(X)_{s_1} \times \cdots \times T_\Sigma(X)_{s_n}$, we will use $T[U] = \bigcup_{\langle t_1, \ldots, t_n \rangle \in U} T(t_1, \ldots, t_n)$.

Given a Σ-algebra $\mathbf{A}$, we will use A_s to denote its carrier set of sort s. As usual, an element of $Hom(\mathbf{T}_\Sigma(\mathbf{X}), \mathbf{A})$ will be called an *assignment* over $\mathbf{A}$. Given the freeness properties of $\mathbf{T}_\Sigma(\mathbf{X})$, an assignment is completely determined by the values it assigns to the elements of X. Given an assignment h over $\mathbf{A}$ and an equation $t_1 \approx t_2$ of sort s, we write $\mathbf{A}, h \Vdash t_1 \approx t_2$ to denote the fact $h(t_1) = h(t_2)$. We say that $\mathbf{A}$ is a model of, or satisfies, an equation $t_1 \approx t_2$ if, for every assignment h over $\mathbf{A}$, we have that $\mathbf{A}, h \Vdash t_1 \approx t_2$. The same applies to conditional equations. Given a class K of Σ-algebras, we define the consequence relation $\models^K_\Sigma$ as follows: $\Theta \models^K_\Sigma t_1 \approx t_2$ when, for every $\mathbf{A} \in K$ and every assignment h over $\mathbf{A}$, if $\mathbf{A}, h \Vdash u_1 \approx u_2$ for every $u_1 \approx u_2 \in \Theta$ then also $\mathbf{A}, h \Vdash t_1 \approx t_2$. We may omit the superscript and simply write $\models_\Sigma$ if K is the class of all Σ-algebras. We will use Eqn^K_Σ to refer to the logic $\langle Eq_\Sigma, \models^K_\Sigma \rangle$.

From now on we will assume that all signatures have a distinguished sort ϕ, for formulas. Moreover, we will assume that $X_\phi = \{\xi_i : i \in \mathbb{N}\}$ and will simply write ξ_k instead of $\xi_k : \phi$. Given a specification $\langle \Sigma, \Phi \rangle$ where Φ is a set of Σ-equations, we define the induced set of formulas $L_{\Sigma,\Phi}$ to be the carrier set of sort ϕ of the initial model $\mathbf{L}_{\Sigma,\Phi} = \mathbf{T}_\Sigma(\emptyset)_{/\Phi}$ of Φ. When $\Phi = \emptyset$, we will simply write $\mathbf{L}_\Sigma$. We define also the set of *schematic formulas* $L_{\Sigma,\Phi}(X)$ to be the carrier set of sort ϕ of the initial model $\mathbf{L}_{\Sigma,\Phi}(\mathbf{X}) = \mathbf{T}_\Sigma(\mathbf{X})_{/\Phi}$ of Φ. When $\Phi = \emptyset$, we will simply write $\mathbf{L}_\Sigma(\mathbf{X})$. Moreover, we will use $BEqn^K_\Sigma$ to refer to the logic $\langle Eq_\Sigma, \models^K_{\Sigma, bhv} \rangle$, where $\models^K_{\Sigma, bhv}$ is the behavioral consequence relation defined for instance as in [5,11], by considering ϕ to be the unique visible sort and adopting a suitable set of visible contexts.

3   Limitations of the Current Theory of AAL

In this section, we intend to illustrate some of the limitations of the current theory of AAL. For that purpose, we begin by briefly presenting the essential notions and results of the theory. Still, it is not our aim to survey AAL, but rather to focus on what will be relevant in the rest of the paper. A recent comprehensive survey of AAL is [2], where the proofs (or pointers to the proofs) of the results we will mention can be found.

3.1   Concepts and Results of Unsorted AAL

The original formulation of AAL in [1] considered only finitary logics. Currently, the finitariness condition has been dropped [2]. Still, the objects of study of current AAL are logics whose formulas have some additional algebraic structure, namely their set of formulas is freely obtained from a propositional (single-sorted) signature.

Definition 1 (Structural single-sorted logic)

A *structural single-sorted logic* is a pair $\mathcal{L} = \langle \Sigma, \vdash \rangle$, where Σ is a single-sorted signature and $\langle L_\Sigma(X), \vdash \rangle$ is a logic that also satisfies:

Structurality: if $\Gamma \vdash \varphi$ then $\sigma[\Gamma] \vdash \sigma(\varphi)$ for every substitution σ.

Clearly, ϕ must be the unique sort of Σ. Note that, if $\langle \Sigma, \vdash \rangle$ is a structural single-sorted logic, then $\langle L_\Sigma, \vdash' \rangle$ is also a logic, where $\vdash'$ is the restriction of $\vdash$ to L_Σ. Finally, we can introduce the main notion of AAL.

Definition 2 (Single-sorted algebraizable logic)

A structural single-sorted logic $\mathcal{L} = \langle \Sigma, \vdash \rangle$ is *algebraizable* if there exists a class K of Σ-algebras, a set $\Theta(\xi)$ of Σ-equations, and a set $E(\xi_1, \xi_2)$ of schematic $\mathcal{L}$-formulas such that the following conditions hold:

- for every $\Gamma \cup \{\varphi\} \subseteq L_\Sigma(X)$, $\Gamma \vdash \varphi$ iff $\Theta[\Gamma] \vDash_\Sigma^K \Theta(\varphi)$;
- for every $\Delta \cup \{\varphi \approx \psi\} \subseteq Eq_\Sigma$, $\Delta \vDash_\Sigma^K \varphi \approx \psi$ iff $E[\Delta] \vdash E(\varphi, \psi)$;
- $\xi \dashv\vdash E[\Theta(\xi)]$ and $\xi_1 \approx \xi_2 \dashv\vDash_\Sigma^K \Theta[E(\xi_1, \xi_2)]$.

The set Θ of equations is called the set of *defining equations*, E is called the set of *equivalential formulas*, and K is called an *equivalent algebraic semantics* for $\mathcal{L}$. The notion of algebraizable logic intuitively means that the consequence relation of $\mathcal{L}$ can be captured by the equational consequence relation $\vDash_K^\Sigma$, and vice-versa, in a logically inverse way. When a logic is algebraizable and both Θ and E are finite, we say the logic is *finitely algebraizable*. Other variants of the notion of algebraizability and their relationships are illustrated in Fig. 1. Note however that, in this paper, we will not explore them.

In [1] Blok and Pigozzi proved interesting results concerning the uniqueness and axiomatization of an equivalent algebraic semantics of a given finitary and finitely algebraizable logic. They proved that a class of algebras K is an equivalent algebraic semantics of a finitary finitely algebraizable logic if and only if the quasivariety generated by K is also an equivalent algebraic semantics. In terms of uniqueness they showed that there is a unique quasivariety equivalent to a given finitary finitely algebraizable logic. The axiomatization of this quasivariety can be directly built from an axiomatization of the logic being algebraized, as stated in the following result.

Theorem 1. *Let $\mathcal{L} = \langle \Sigma, \vdash \rangle$ be a (finitary) structural single-sorted logic obtained from a deductive system formed by a set of axioms Ax and a set of inference rules R. Assume that $\mathcal{L}$ is finitely algebraizable with Θ and E. Then, the*

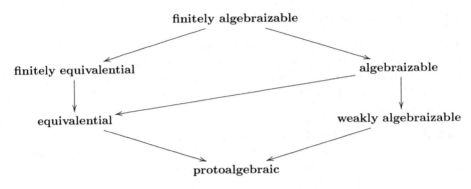

Fig. 1. A view of the Leibniz hierarchy

equivalent quasivariety semantics is axiomatized by the following equations and conditional-equations:

- $\Theta(\varphi)$ *for each* $\varphi \in Ax$;
- $\Theta[E(\xi, \xi)]$;
- $\Theta(\psi_0) \wedge \ldots \wedge \Theta(\psi_n) \to \Theta(\psi)$ *for each rule* $\frac{\psi_0 \ldots \psi_n}{\psi} \in R$;
- $\Theta[E(\xi_1, \xi_2)] \to \xi_1 \approx \xi_2$.

There are several interesting and useful alternative characterizations of algebraizability. The most useful, namely to prove negative results, is perhaps the characterization that explores the properties of the so-called *Leibniz operator*. A congruence $\equiv$ in a Σ-algebra **A** is said to be *compatible* with a subset F of A_ϕ if whenever $a \in F$ and $a \equiv b$ then $b \in F$. In this case, F is a union of equivalence classes of $\equiv$. We will use $\mathrm{Cong_A}$ to denote the set of all congruences of a Σ-algebra **A**. Recall that $\mathrm{Cong_A}$ equipped with inclusion also constitutes a complete partial order.

Definition 3 (Leibniz operator)
Let $\mathcal{L} = \langle \Sigma, \vdash \rangle$ be a structural single-sorted logic. The *Leibniz operator* on the formula algebra, $\Omega : Th_{\mathcal{L}} \to \mathrm{Cong}_{\mathbf{L_\Sigma(X)}}$ is such that, for each theory Γ of $\mathcal{L}$, $\Omega(\Gamma)$ is the largest congruence of $\mathbf{L_\Sigma(X)}$ compatible with Γ.

The denomination of the hierarchy considered in Fig. 1 is well justified by the fact that each of the classes of logics mentioned can be characterized by inspection of the properties of the corresponding Leibniz operator. Concerning algebraizability, we have the following result.

Theorem 2. *A structural single-sorted logic* $\mathcal{L} = \langle \Sigma, \vdash \rangle$ *is algebraizable iff* Ω *is monotone, injective, and commutes with inverse substitutions.*

Another enlightening characterization of algebraization can be expressed using maps of logics [12].

Theorem 3. *A structural single-sorted logic* $\mathcal{L} = \langle \Sigma, \vdash \rangle$ *is algebraizable iff there exists a class K of Σ-algebras and a strong representation (θ, τ) of $\mathcal{L}$ in Eqn_K^Σ such that both θ and τ commute with substitutions.*

Note that the fact that both maps commute with substitutions is essential to guarantee that each can be given uniformly, respectively, by a set Θ of one-variable equations, and a set E of two-variable formulas.

3.2 Examples, Good and Bad

The theory of AAL is fruitful in positive and interesting examples. We will begin by introducing two well known simple examples.

Example 1 (Classical and Intuitionistic Propositional Logics)
The main paradigm of AAL is the well establish connection between *classical propositional logic (CPL)* and the variety of *Boolean algebras*. This was really the starting point to the idea of connecting logic with algebra, which evolved trying to generalize this connection to other logics. Another important example is *intuitionistic propositional logic (IPL)*. Its algebraization gives rise to the class of *Heyting algebras*. It is interesting to note that, in contrast to Boolean algebras, which were defined before the Lindenbaum-Tarski techniques were first applied to generate them from CPL, Heyting algebras seem to be the first algebras of logic that were identified precisely by applying these techniques (which are the ancient roots of the modern theory of AAL) to a given axiomatization of IPL.

Even considering the enormous success of this theory, not only in the generality of its results, but also in the large amount of examples, we can point out some limitations. From our point of view, one major limitation of the existing theory is its inability to correctly deal with logics with a many-sorted language. Let us, first of all, discuss the paradigmatic example of *first-order classical logic (FOL)*.

Example 2 (First-Order Classical Logic)
Research on the algebraization of FOL goes back to the seminal work initiated by Tarski in the 1940s, and published in collaboration with Henkin and Monk in [13]. This line of research is known as the *cylindric* approach, the one that we will follow here. Nevertheless, we mention that there is another important approach to the algebraization of FOL, known as the *polyadic* approach, that differs from the cylindric approach mainly because it deals with explicit substitutions.

In [1], Blok and Pigozzi, following the cylindric approach, presented a single-sorted algebraization of FOL in the terms we have just introduced. Their idea was to massage the first-order language into a propositional language and then present a structural propositional deductive system **PR**, introduced by Németi, for first-order logic over this propositional language. It is then proved that **PR** is algebraizable. Moreover, it is proved that the variety equivalent to **PR** is the variety of cylindric algebras.

Despite the success of the example of FOL within AAL, we can point out some drawbacks. The first one is related to the fact that the first-order language they start from differs in several important respects from standard FOL. This is due to the fact that, since the theory only applies to single-sorted logics, the atomic formulas of FOL have to be represented, within the propositional language, as propositional variables. So, in order to preserve structurality, one has to constrain the language.

Another important drawback is the fact that, given the many-sorted character of first-order logic, where we have at least syntactic categories for terms and formulas, and possibly also for variables, it would be desirable to have an algebraic counterpart that reflects this many-sorted character. This is clearly not the case with cylindric algebras.

One of our motivations is precisely to extend the theory of AAL to cope with logics that, like first-order logic, have a many-sorted language. This will allow us to, given an algebraizable many-sorted logic, reflect its many-sorted character in its corresponding algebraic counterpart.

But it is not only at the purely many-sorted level that the limitations of current AAL arise. Even at the propositional level, there are interesting logics that fall out of the scope of the theory. It is the case of certain so-called *non-truth-functional logics*, such as the paraconsistent systems of da Costa [7]. The major problem with these logics is that they lack congruence for some connective(s). Roughly speaking, a logic is said to be paraconsistent if its consequence relation is not *explosive* [14]. We say that a logic $\mathcal{L} = \langle \Sigma, \vdash \rangle$ is an *explosive logic* with respect to a *negation* connective $\neg$ if, for all formulas φ and ψ, it is true that $\{\varphi, \neg\varphi\} \vdash \psi$.

Example 3 (Paraconsistent Logic $\mathcal{C}_1$ of da Costa)
It was proved, first by Mortensen [8], and after by Lewin, Mikenberg and Schwarze [9] that $\mathcal{C}_1$ is not algebraizable in current AAL. So, we can say that $\mathcal{C}_1$ is an example of a logic whose non-algebraizability is well studied. Nevertheless, it is rather strange that a relatively well-behaved logic fails to have an algebraic counterpart. We will briefly introduce $\mathcal{C}_1$.

The language of $\mathcal{C}_1$ is generated by the unisorted signature Σ with sort ϕ and composed of the following operations:

- $\mathbf{t}, \mathbf{f} :\to \phi$, $\neg : \phi \to \phi$ and $\wedge, \vee, \supset: \phi^2 \to \phi$.

The consequence relation of $\mathcal{C}_1$ can be given by the structural deductive system composed of the following axioms:

- $\xi_1 \supset (\xi_2 \supset \xi_1)$
- $(\xi_1 \supset (\xi_2 \supset \xi_3)) \supset ((\xi_1 \supset \xi_2) \supset (\xi_1 \supset \xi_3))$
- $(\xi_1 \wedge \xi_2) \supset \xi_1$
- $(\xi_1 \wedge \xi_2) \supset \xi_2$
- $\xi_1 \supset (\xi_2 \supset (\xi_1 \wedge \xi_2))$
- $\xi_1 \supset (\xi_1 \vee \xi_2)$
- $\xi_2 \supset (\xi_1 \vee \xi_2)$
- $(\xi_1 \supset \xi_3) \supset ((\xi_2 \supset \xi_3) \supset ((\xi_1 \vee \xi_2) \supset \xi_3))$
- $\neg\neg\xi_1 \supset \xi_1$
- $\xi_1 \vee \neg\xi_1$
- $\xi_1^\circ \supset (\xi_1 \supset (\neg\xi_1 \supset \xi_2))$

- $(\xi_1^\circ \wedge \xi_2^\circ) \supset (\xi_1 \wedge \xi_2)^\circ$
- $(\xi_1^\circ \wedge \xi_2^\circ) \supset (\xi_1 \vee \xi_2)^\circ$
- $(\xi_1^\circ \wedge \xi_2^\circ) \supset (\xi_1 \supset \xi_2)^\circ$
- $\mathbf{t} \equiv (\xi_1 \supset \xi_1)$
- $\mathbf{f} \equiv (\xi_1^\circ \wedge (\xi_1 \wedge \neg\xi_1))$

and the rule of *modus ponens*:

- $\xi_1, \xi_1 \supset \xi_2 \vdash \xi_2$

where φ° is an abbreviation of $\neg(\varphi \wedge (\neg\varphi))$ and $\varphi \equiv \psi$ is an abbreviation of $(\varphi \supset \psi) \wedge (\psi \supset \varphi)$.

Despite of its innocent aspect, $\mathcal{C}_1$ is a non-truth-functional logic, namely it lacks congruence for its paraconsistent negation connective. In general, it may happen that $\varphi \dashv\vdash \psi$ but $\neg\varphi \not\dashv\vdash \neg\psi$. This phenomenon leaves the non-truth-functional logics, and in particular $\mathcal{C}_1$, outside of the existing theory of AAL, since congruence is a key ingredient for the algebraization process. Still, $\mathcal{C}_1$ has other peculiarities. Although it is defined as a logic weaker than CPL, it happens that a classical negation $\sim$ can be defined in $\mathcal{C}_1$ by using the abbreviation $\sim \varphi = \varphi^\circ \wedge \neg\varphi$. Exploring this fact, da Costa himself introduced in [15] a so-called class of *Curry algebraic structures* as a possible algebraic counterpart of $\mathcal{C}_1$. In fact, nowadays, these algebraic structures are known as *da Costa algebras* [16]. However, their precise nature remains unknown, given the non-algebraizability results reported above.

One of the objectives of this paper is to point out a way to use our many-sorted approach to give a possible connection between $\mathcal{C}_1$ and the algebras of da Costa.

4 Generalizing Algebraization

In this section we will propose a novel many-sorted extension of the notion of algebraization, where the major generalizations will happen at the syntactic level. The new notion is then illustrated with the help of examples, and some of its essential results are extended to the many-sorted setting.

4.1 The Many-Sorted Generalization

Our initial aim is to extend the range of applicability of AAL. Therefore, we need to introduce a suitable notion of *structural many-sorted logic*. First of all, we will assume that the formulas of the logic are built from a many-sorted signature Σ. It is usual to assume that the syntax, namely of a logic, is defined by a free construction over the given signature Σ. In the previous section the set of formulas was precisely T_Σ, as inforced by the terminology introduced in section 2 when the formulas are built from a specification $\langle \Sigma, \Phi \rangle$ with $\Phi = \emptyset$. However, it is not unusual that certain syntactic abbreviations are assumed. For instance, in CPL one may assume that all classical connectives are primitive, or else, for instance, that negation and implication are primitive and the other connectives appear as abbreviations. We have also the example of α-congruence in the λ-calculus. In such a scenario, it makes all the sense to assume that these syntactic abbreviations correspond to equations over the syntax, thus making $\Phi \neq \emptyset$. Of course, by doing this one may be contributing to blurring the essential distinction between syntax and semantics. Still, as we will see, the development applies unrestrictedly. Hence, in general, the syntax of the logic will be specified by a pair $\langle \Sigma, \Phi \rangle$, which justifies the general definition of $L_{\Sigma,\Phi}$. Note that given a substitution σ over Σ, it is easy to see that σ is well behaved with respect to the congruence $\equiv_\Phi$ induced by Φ on $\mathbf{L}_\Sigma(\mathbf{X})$. Namely, given $t_1, t_2 \in L_\Sigma(X)$ if $t_1 \equiv_\Phi t_2$ then $\sigma(t_1) \equiv_\Phi \sigma(t_2)$. Hence, it makes sense to write $\sigma(\varphi)$ for any schema formula $\varphi \in L_{\Sigma,\Phi}(X)$. We will use $[t]_\Phi$ to denote the schema formula corresponding to the equivalence class of $t \in L_\Sigma(X)$ under $\equiv_\Phi$.

Definition 4 (Structural many-sorted logic)
A *structural many-sorted logic* is a tuple $\mathcal{L} = \langle \Sigma, \Phi, \vdash \rangle$ where $\langle \Sigma, \Phi \rangle$ is a many-sorted specification and $\langle L_{\Sigma,\Phi}(X), \vdash \rangle$ is a logic that also satisfies:

Structurality: if $\Gamma \vdash \varphi$ then $\sigma[\Gamma] \vdash \sigma(\varphi)$ for every substitution σ.

It should be clear that, in the particular case of a single-sorted signature with $\Phi = \emptyset$, this notion coincides precisely with the notion of structural single-sorted logic used in the previous section. Note also that, given a structural many-sorted logic $\mathcal{L} = \langle \Sigma, \Phi, \vdash \rangle$, we can consider the following induced consequence relation $\vdash_{\mathcal{L}}^{\Phi} \subseteq \mathcal{P}(L_\Sigma(X)) \times L_\Sigma(X)$ defined by $T \vdash_{\mathcal{L}}^{\Phi} u$ iff $[T]_\Phi \vdash [u]_\Phi$. It is easy to see that there exists a strong representation of $\mathcal{L}$ into $\langle L_\Sigma(X), \vdash_{\mathcal{L}}^{\Phi} \rangle$. In particular, the theories $Th_{\mathcal{L}}^{\Phi}$ of this induced logic are isomorphic to $Th_{\mathcal{L}}$. When convenient, this induced consequence relation will allow us to work over $L_\Sigma(X)$, thus avoiding the explicit reference to quotients and equivalence classes. We can now introduce our new notion of many-sorted algebraization. The key idea is to replace the role of single-sorted equational logic in current AAL by many-sorted equational logic.

Definition 5 (Many-sorted algebraizable logic)
A structural many-sorted logic $\mathcal{L} = \langle \Sigma, \Phi, \vdash \rangle$ is algebraizable if there exists a class K of $\langle \Sigma, \Phi \rangle$-models, a set $\Theta(\xi)$ of ϕ-equations and a set $E(\xi_1, \xi_2)$ of ϕ-terms such that the following conditions hold:

  i) for every $T \cup \{u\} \subseteq L_\Sigma(X)$, $T \vdash_{\mathcal{L}}^{\Phi} u$ iff $\Theta[T] \vDash_\Sigma^K \Theta(u)$;

  ii) for every set $\Delta \cup \{t_1 \approx t_2\}$ of ϕ-equations, $\Delta \vDash_\Sigma^K t_1 \approx t_2$ iff $E[\Delta] \vdash_{\mathcal{L}}^{\Phi} E(t_1, t_2)$;

  iii) $\xi \dashv\vdash_{\mathcal{L}}^{\Phi} E[\Theta(\xi)]]$ and $\xi_1 \approx \xi_2 \dashv\vDash_\Sigma^K \Theta[E(\xi_1, \xi_2)]$.

As before, Θ is called the set of defining equations, E the set of equivalential formulas, and K is called an equivalent algebraic semantics for $\mathcal{L}$. Again, it should be clear that in the case of a single-sorted signature with $\Phi = \emptyset$ this definition coincides with the notion of algebraizable logic of current AAL.

4.2   Examples

Before we proceed, let us illustrate the new notion, namely by revisiting the examples of FOL and C_1.

Example 4 (First-order Classical Logic Revisited)
In example 2, we have already discussed the problems with the single-sorted algebraization of FOL developed in [1]. With our many-sorted framework we can now handle first-order logic as a two-sorted logic, with a sort for terms and a sort for formulas. This perspective seems to be much more convenient, and we no longer need to view atomic FOL formulas as propositional variables. Working out the example, whose details we omit, we manage to algebraize FOL having as an equivalent algebraic semantics the class of *two-sorted cylindric algebras*, whose restriction to the sort ϕ is a plain old cylindric algebra, but which corresponds to a regular first-order interpretation structure on the sort of terms. In the new many-sorted context it is also straightforward to algebraize many-sorted FOL.

Example 5 (C_1 Revisited)
In example 3, we made clear that the single-sorted theory of AAL has some unexpected limitations, even in the case of propositional-based logics. We will now revisit C_1 and its algebraization in the many-sorted setting. Actually, in this new perspective, the way in which a logic is presented and, in particular, the way its language is specified, is very relevant in the algebraization process. The trick for C_1 will be to present it as a two-sorted logic, as suggested in [17]. Namely we will consider the two-sorted syntactic specification $\langle \Sigma, \phi \rangle$ such that:

- Σ has two sorts, h and ϕ, and operations $\mathbf{t}, \mathbf{f} :\to h$, $\neg, \sim: h \to h$ and $\wedge, \vee, \supset: h^2 \to h$, as well as $o : h \to \phi$, and $\mathbf{t}, \mathbf{f} :\to \phi$, $\sim: \phi \to \phi$ and $\wedge, \vee, \supset: \phi^2 \to \phi$;
- Φ includes the following equations:

$$\sim x \approx x^{\circ} \wedge \neg x$$

$$o(\mathbf{t}) \approx \mathbf{t} \qquad o(\sim x) \approx \sim o(x) \qquad o(x \wedge y) \approx o(x) \wedge o(y)$$
$$o(\mathbf{f}) \approx \mathbf{f} \qquad o(x \vee y) \approx o(x) \vee o(y) \qquad o(x \supset y) \approx o(x) \supset o(y)$$

where x and y are variables of sort h. The idea is to take all the primitive syntax of C_1 to the sort h, including the classical negation connective $\sim$ definable as an abbreviation, and to have an observation operation o into sort ϕ, where all the connectives are again available, with the exception of the non-truth-functional paraconsistent negation $\neg$. The top equation aims precisely at internalizing the definition of $\sim$. The other 6 equations simply express the truth-functional (homomorphic) nature of the corresponding connectives. It is not difficult to see that $L_{\Sigma,\Phi}$ is isomorphic to the set of C_1-formulas. With this two-sorted perspective, it can now be shown that, taking $\Theta(\xi) = \{\xi \approx \mathbf{t}\}$ as the set of defining equations and $E(\xi_1, \xi_2) = \{\xi_1 \equiv \xi_2\}$ as the set of equivalential formulas, C_1 is algebraizable in our generalized sense, and that the resulting algebraic counterpart is precisely the two-sorted quasivariety K_{C_1} proposed in [17]. We do not dwell into the details here, but we can say that the corresponding two-sorted algebras are Boolean on sort ϕ. Actually, the conditional equational specification of K_{C_1}

only needs to use ϕ-equations, which leaves little to be said about what happens with sort h. It is certainly very interesting to understand what is the impact of the $K_{\mathcal{C}_1}$ specification over h-terms, but that is something that we can only do behaviorally, by assuming that h is a hidden-sort. If we restrict our attention to contexts that do not involve the paraconsistent negation, we can show that every algebra $\mathbf{A} \in K_{\mathcal{C}_1}$ behaviorally satisfies all the conditions in the definition of da Costa algebras. On the other hand, given any da Costa algebra, we can canonically extend it to a two-sorted algebra in $K_{\mathcal{C}_1}$. In this way, we manage to discover the connection of da Costa algebras with the algebraization of $\mathcal{C}_1$, which had never been found.

4.3 Many-Sorted AAL

In order to further support our generalization of the notion of algebraizable logic, we will now show that we can also extend other notions and results of AAL. We begin by defining a many-sorted version of the Leibniz operator.

Definition 6 (Many-sorted Leibniz operator)

Let $\mathcal{L} = \langle \Sigma, \Phi, \vdash \rangle$ be a structural many-sorted logic. The *many-sorted Leibniz operator* on the term algebra, $\Omega : Th_{\mathcal{L}}^{\Phi} \to Cong_{\mathbf{L}_{\Sigma}(\mathbf{X})}$ is such that, for each $T \in Th_{\mathcal{L}}^{\Phi}$, $\Omega(T)$ is the largest congruence of $\mathbf{L}_{\Sigma}(\mathbf{X})$ containing Φ and compatible with T.

Note that, given $T \in Th_{\mathcal{L}}^{\Phi}$, since $\Phi \subseteq \Omega(T)$, we have that $\Omega(T)$ can be seen as a congruence on $\mathbf{L}_{\Sigma}(\mathbf{X})_{/\phi}$. As we will see, also in the many-sorted setting, the Leibniz operator will play an important role. In fact, we are able to generalize the characterization theorem of single-sorted algebraizable logic we gave in section 3.1.

Theorem 4. *A structural many-sorted logic $\mathcal{L} = \langle \Sigma, \Phi, \vdash \rangle$ is algebraizable iff Ω is monotone, injective, and commutes with inverse substitutions.*

Proof. This proof uses the same methodology as the proof of the single-sorted result. So, we will just give a sketch of the proof focusing on the important methodological steps. First assume that $\mathcal{L}$ is algebraizable, with K, $\Theta(\xi)$ and $E(\xi_1, \xi_2)$. Using $\Theta(\xi)$ and its properties we can define the function $\Omega_K : Th_{\mathcal{L}}^{\Phi} \to Cong_{\mathbf{L}_{\Sigma}(\mathbf{X})}$, such that, for every sort s, $\langle t_1, t_2 \rangle \in (\Omega_K(T))_s$ iff for every ϕ-term $u(x : s)$ we have that $\Theta[T] \vDash_{\mathcal{L}}^{K} u(t_1) \approx u(t_2)$. Using the properties of K, $\Theta(\xi)$ and $E(\xi_1, \xi_2)$ it is easy to prove that $\Omega_K(T)$ is the largest congruence containing Φ that is compatible with T, that is, $\Omega_K = \Omega$. The fact that Ω_K is injective, monotone and commutes with inverse substitutions also follows easily from the properties of $\Theta(\xi)$ and $E(\xi_1, \xi_2)$.

On the other direction, suppose that Ω is injective, monotone and commutes with inverse substitutions. Consider the class of algebras $K = \{ \mathbf{T}_{\Sigma}(\mathbf{X})_{/\Omega(T)} : T \in Th_{\mathcal{L}}^{\Phi} \}$. It is clear that K is a class of $\langle \Sigma, \Phi \rangle$-models. The fact that Ω is monotone and commutes with inverse substitutions implies, according to [18], that Ω is also surjective. Hence, Ω is indeed a bijection. Our objective is to prove that $\mathcal{L}$ is algebraizable with K an equivalent algebraic semantics. We still have to find the sets of defining equations and equivalence formulas. Let $T = \{\xi_1\}^{\vdash_{\mathcal{L}}^{\Phi}}$. Let σ be the substitution over Σ such that $\sigma_\phi(\xi) = \xi_1$ for every ξ, and it is the identity in the other sorts. If we take $\Theta = \sigma[\Omega(T)_\phi]$ it can be shown, using the fact that Ω commutes with inverse substitutions, that $\Theta = \Theta(\xi_1)$ is a set of ϕ-equations and $T \vdash_{\mathcal{L}}^{\Phi} u$ iff $\Theta[T] \vDash_{\Sigma}^{K} \Theta(u)$. Now let us construct the set of equivalence formulas. Take now σ to be the substitution such that $\sigma_\phi(\xi_2) = \xi_2$ and $\sigma_\phi(\xi) = \xi_1$ for every $\xi \neq \xi_2$, and it is the identity in the other sorts. Take $E = \sigma[\Omega^{-1}(\{\xi_1 \approx \xi_2\}^{\vDash_{\Sigma}^{K}})]$. It can be proved that $E = E(\xi_1, \xi_2)$ is a set of ϕ-terms and that $\Theta[E(\xi_1, \xi_2)] =\!\vDash_{\Sigma}^{K} \xi_1 \approx \xi_2$. The algebraizability of $\mathcal{L}$ follows straightforwardly from these facts. □

We can also extend the characterization of algebraization using maps of logics, namely a strong representation between the given many-sorted logic and many-sorted equational logic.

Theorem 5. *A structural many-sorted logic $\mathcal{L} = \langle \Sigma, \Phi, \vdash \rangle$ is algebraizable iff there exists a class K of $\langle \Sigma, \Phi \rangle$-models and a strong representation $\langle \theta, \tau \rangle$ of $\langle L_\Sigma(X), \vdash_{\mathcal{L}}^{\Phi} \rangle$ in Eqn_Σ^K, such that θ is given by a set $\Theta(\xi)$ of ϕ-equations and τ by a set $E(\xi_1, \xi_2)$ of ϕ-terms.*

Proof. The result follows from the observation that conditions i), ii), and iii) of the definition of many-sorted algebraizable logic are equivalent to the fact that $\langle \theta, \tau \rangle$ is a strong representation. □

When a structural many-sorted logic $\mathcal{L}$ is algebraizable, we can sometimes provide a specification of its algebraic counterpart given a deductive system for $\mathcal{L}$.

Theorem 6. *Let $\mathcal{L} = \langle \Sigma, \Phi, \vdash \rangle$ be a structural many-sorted logic obtained from a deductive system formed by a set Ax of axioms and a set R of inference rules. Assume that $\mathcal{L}$ is finitely algebraizable with Θ and E. Then, the equivalent quasivariety semantics is axiomatized by the following equations and conditional-equations:*

    i) Φ;
    ii) $\Theta(\varphi)$ *for each* $[\varphi]_\Phi \in Ax$;
    iii) $\Theta[E(\xi, \xi)]$;
    iv) $\Theta(\psi_0) \wedge \ldots \wedge \Theta(\psi_n) \to \Theta(\psi)$ *for each* $\frac{[\psi_0]_\Phi, \ldots, [\psi_n]_\Phi}{[\psi]_\Phi} \in R$;
    v) $\Theta[E(\xi_1, \xi_2)] \to \xi_1 \approx \xi_2$.

Proof. Let K be the quasivariety defined by *i)-v)*. We will prove that K is the equivalent algebraic semantics of $\mathcal{L}$. First note that the fact that K satisfies *i)* is equivalent to the fact that K is a class of $\langle \Sigma, \Phi \rangle$-models. It is easy to prove that equation *iii)* and conditional equation *v)* are jointly equivalent to $\xi_1 \approx \xi_2 =\!\vDash_{\Sigma}^{K} \Theta[E(\xi_1, \xi_2)]$ which is one-half of condition iii) in the definition of many-sorted algebraizable logic. It can also be verified that condition i) of the definition of algebraizable logic is equivalent to the above equations *ii)* and *iv)*.

It now remains to say that, as it is well known in the single-sorted case, this is enough to guarantee the algebraizability of $\mathcal{L}$. The uniqueness of the equivalent quasivariety in this case is straightforward.   □

Note that, due to the inclusion of the equations in Φ, items $ii)$ and $iv)$ are independent of the particular choice of representatives of the equivalence classes.

At this point it is important to clarify the precise role of behavioral logic in our framework. It should be clear that we do not use behavioral reasoning directly in the algebraization process, since this is built over plain old many-sorted equational logic. However, following the spirit of hidden equational logic, and considering the sort of formulas as the only visible sort, we might think of behaviorally reasoning about the other sorts. It is exactly here that we use behavioral logic to reason about the algebraic counterpart of a given algebraizable logic. This possibility was particularly useful in example 5.

It is well known that behavioral reasoning can be a very complex issue, since it involves reasoning about all possible experiments we can perform on a hidden term. There are, nevertheless, some nice approaches to tackle this problem [11,5]. Here, we identify a very particular case that occurs when the behavioral reasoning associated with a class K of algebras is *specifiable*, in the sense that all the behaviorally valid equations can be derived, in standard equational logic, from some set of (possibly hidden) equations. We will show that, when $\mathcal{L}$ is expressive enough and it is finitely algebraizable, the behavioral reasoning (over the whole signature) associated with an equivalent algebraic semantics for $\mathcal{L}$ is specifiable. A structural many-sorted logic $\mathcal{L} = \langle \Sigma, \Phi, \vdash \rangle$ is said to be *observationally equivalential* if there exists a sorted set $E = \{E_s(x_1 : s, x_2 : s)\}_{s \in S}$ of ϕ-terms such that, for every $s \in S$ and $x, y, z \in X_s$:

- $\vdash^\Phi_\mathcal{L} E_s(x, x)$;
- $E_s(x, y) \vdash^\Phi_\mathcal{L} E_s(y, x)$;
- $E_s(x, y), E_s(y, z) \vdash^\Phi_\mathcal{L} E_s(x, z)$;
- for each operation $o \in O_{s_1 \ldots s_n s}$ and $x_1, y_1 \in X_1, \ldots, x_n, y_n \in X_n$
  $\{E_{s_1}(x_1, y_1), \ldots, E_{s_n}(x_n, y_n)\} \vdash^\Phi_\mathcal{L} E_s(o(x_1, \ldots, x_n), o(y_1, \ldots, y_n))$;
- $E_\phi(\xi_1, \xi_2), \xi_1 \vdash^\Phi_\mathcal{L} \xi_2$.

Each E_s is called the set of *equivalential terms* of sort s. If, for each $s \in S$, the set E_s is finite, then $\mathcal{L}$ is called finitely observationally equivalential.

Theorem 7. *Let $\mathcal{L} = \langle \Sigma, \Phi, \vdash \rangle$ be a finitely algebraizable, finitary, and structural many-sorted logic, and K be an equivalent algebraic semantics for $\mathcal{L}$. If $\mathcal{L}$ is finitely observationally equivalential then $\models^K_{\Sigma, bhv}$ is specifiable.*

Proof. Since $\mathcal{L}$ is finitary and finitely algebraizable, $\models^K_\Sigma$ is specifiable. Let $E = \{E_s\}_{s \in S}$ the S-indexed set of equivalential formulas. Consider the sorted set $\Psi = \{\Psi_s\}_{s \in S}$ such that $\Psi_s = \Theta[E_s]$, where $\Theta(\xi)$ is the (finite) set of defining equations. Then, it is easy to prove that Ψ forms a finite set of equivalential formulas for $\models^K_\Sigma$. Then, using theorem 5.2.21. in [19], $\models^K_{\Sigma, bhv}$ is specifiable.   □

Note that this theorem does not apply to $\mathcal{C}_1$, in examples 3 and 5, since $\mathcal{C}_1$ is not observationally equivalential due to the non-congruence of its paraconsistent negation. This fact motivated the idea, in example 5, of not considering paraconsistent negation as a visible connective. The fact that the other connectives remain visible can be justified by a maximality argument. Here we point out the importance of investigating general results relating algebraization with the choice of visible connectives, namely when applying this framework to other non-truth-functional logics.

5   Conclusions and Further Work

In this paper, we have proposed a generalization of the notion of algebraizable logic that encompasses also many-sorted logics. The key ingredient of the generalization was to replace the role of single-sorted equational logic of traditional AAL by many-sorted equational logic. To support our approach we proved, in this more general setting, extended versions of several important results of AAL, including characterizations using the Leibniz operator, as well as maps of logics. We illustrated the approach by reanalyzing the examples of first-order logic and of the paraconsistent logic $\mathcal{C}_1$ in a many-sorted context. In particular, for $\mathcal{C}_1$, we managed to characterize the precise role of da Costa algebras.

Being a first attempt at this generalization, there is much to be done, and there are many more interesting results of AAL to generalize. We also aim at investigating a many-sorted version of the Leibniz hierarchy, including also protoalgebraization, weak-algebraization, and related work, such as k-deductive systems. Many further examples are also to be tried. In this front, our ultimate aim is to understand the relationship between orthomodular lattices as used in the Birkhoff and Von Neumann tradition of quantum logic and the algebraic counterpart of exogenous quantum logic [20]. An important open question is whether and how our approach can be integrated with the work on the algebraization of logics as institutions reported in [21]. Another interesting line of future work is to study the impact of our proposal with respect to the way a logic is represented within many-sorted equational logic in the context of logic combination, namely in the lines of [3].

We already have some ideas on how to incorporate behavioral reasoning directly in the algebraization process. This will allow for a better treatment of non-truth-functional logics and will be the subject of a forthcoming paper. In any case, this seems to be an area which is very fit for application of the theory of many-sorted algebras, including hidden-sorts and behavioral reasoning, as developed within the formal methods community over the last couple of decades.

References

1. Blok, W., Pigozzi, D.: Algebraizable logics. Memoirs of the AMS **77** (1989)
2. Font, J., Jansana, R., Pigozzi, D.: A survey of abstract algebraic logic. Studia Logica **74** (2003) 13–97

3. Mossakowski, T., Tarlecki, A., Pawłowski, W.: Combining and representing logical systems using model-theoretic parchments. In: Recent Trends in Algebraic Development Techniques. Volume 1376 of LNCS. Springer-Verlag (1998) 349–364
4. Sernadas, A., Sernadas, C., Caleiro, C.: Fibring of logics as a categorial construction. Journal of Logic and Computation **9** (1999) 149–179
5. Roşu, G.: Hidden Logic. PhD thesis, University of California at San Diego (2000)
6. Blok, W., Pigozzi, D.: Algebraic semantics for universal Horn logic without equality. In: Universal algebra and quasigroup theory, Lect. Conf., Jadwisin/Pol. 1989. Volume 19. (1992) 1–56
7. da Costa, N.: On the theory of inconsistent formal systems. Notre Dame Journal of Formal Logic **15** (1974) 497–510
8. Mortensen, C.: Every quotient algebra for C_1 is trivial. Notre Dame Journal of Formal Logic **21** (1980) 694–700
9. Lewin, R., Mikenberg, I., Schwarze, M.: C_1 is not algebraizable. Notre Dame Journal of Formal Logic **32** (1991) 609–611
10. Wójcicki, R.: Theory of Logical Calculi. Synthese Library. Kluwer Academic Publishers (1988)
11. Goguen, J., Malcolm, G.: A hidden agenda. Theoretical Computer Science **245** (2000) 55–101
12. Czelakowski, J.: Protoalgebraic logics. Volume 10 of Trends in Logic—Studia Logica Library. Kluwer Academic Publishers (2001)
13. Henkin, L., Monk, J.D., Tarski, A.: Cylindric algebras. Studies in Logic. North-Holland Publishing Company (1971)
14. Carnielli, W.A., Marcos, J.: A taxonomy of C-systems. In Carnielli, W.A., Coniglio, M.E., D'Ottaviano, I.M.L., eds.: Paraconsistency: The logical way to the inconsistent. Volume 228 of Lecture Notes in Pure and Applied Mathematics. (2002) 1–94
15. da Costa, N.: Opérations non monotones dans les treillis. Comptes Rendus de l'Academie de Sciences de Paris (1966) A423–A429
16. Carnielli, W.A., de Alcantara, L.P.: Paraconsistent algebras. Studia Logica (1984) 79–88
17. Caleiro, C., Carnielli, W., Coniglio, M., Sernadas, A., Sernadas, C.: Fibring non-truth-functional logics: Completeness preservation. Journal of Logic, Language and Information **12** (2003) 183–211
18. Herrmann, B.: Characterizing equivalential and algebraizable logics by the Leibniz operator. Studia Logica (1996) 419–436
19. Martins, M.: Behavioral reasoning in generalized hidden logics. Phd Thesis. Faculdade de Ciências, University of Lisbon (2004)
20. Mateus, P., Sernadas, A.: Weakly complete axiomatization of exogenous quantum propositional logic. Information and Computation **204** (2006) 771–794 ArXiv math.LO/0503453.
21. Voutsadakis, G.: Categorical abstract algebraic logic: algebraizable institutions. In: Applied Categorical Structures. Volume 10 (6). Springer-Verlag (2002) 531–568

Algebraic Semantics of Service Component Modules*

José Luiz Fiadeiro[1], Antónia Lopes[2], and Laura Bocchi[1]

[1] Department of Computer Science, University of Leicester
University Road, Leicester LE1 7RH, UK
{bocchi,jose}@mcs.le.ac.uk
[2] Department of Informatics, Faculty of Sciences, University of Lisbon
Campo Grande, 1749-016 Lisboa, Portugal
mal@di.fc.ul.pt

Abstract. We present a notion of module acquired from developing an algebraic framework for service-oriented modelling. More specifically, we give an account of the notion of module that supports the composition model of the SENSORIA Reference Modelling Language (SRML). The proposed notion is independent of the logic in which properties are expressed and components are programmed. Modules in SRML are inspired in concepts proposed for Service Component Architecture (SCA) and Web Services, as well the modules that have been proposed for Algebraic Specifications, namely by H. Ehrig and F. Orejas, among others; they include interfaces for required (imported) and provided (exported) services, as well as a number of components (body) whose orchestrations ensure how given behavioural properties of the provided services are guaranteed assuming that the requested services satisfy required properties.

1 Introduction

In the emerging service-oriented computing paradigm, services are understood as autonomous, platform-independent computational entities that can be described, published, discovered, and dynamically assembled for developing massively distributed, interoperable, evolvable systems. In order to cope with the levels of complexity entailed by this paradigm, one needs abstractions through which complex systems can be understood in terms of compositions of simpler units that capture structures of the application domain. This is why, within the IST-FET Integrated Project SENSORIA – *Software Engineering for Service-Oriented Overlay Computers* – we are developing an algebraic framework for supporting service-oriented modelling at levels of abstraction that are closer to the "business domain".

More precisely, we are defining a suite of languages that support different activities in service-oriented modelling to be adopted as a reference modelling "language" – SRML – within the SENSORIA project. In this paper, we are concerned with the "composition language" SRML-P through which service compositions can be modelled in the form of business processes, independently of the hosting middleware

---

* This work was partially supported through the IST-2005-16004 Integrated Project *SENSORIA: Software Engineering for Service-Oriented Overlay Computers*, and the Marie-Curie TOK-IAP MTK1-CT-2004-003169 *Leg2Net: From Legacy Systems to Services in the Net.*

J.L. Fiadeiro and P.-Y. Schobbens (Eds.): WADT 2006, LNCS 4409, pp. 37–55, 2007.
© Springer-Verlag Berlin Heidelberg 2007

and hardware platforms, and the languages in which services are programmed. The cornerstone of this language is the notion of *module* through which one can model composite services understood as services whose business logic involves the invocation of other services.

In our approach, a module captures a business process that interacts with a set of external services to achieve a certain "goal". This goal should not be understood as a "return value" to be achieved by a computation in the traditional sense, but as a "business interaction" that is offered for other modules to discover and engage with. Global business goals emerge not from prescribed computations but from the peer-to-peer, conversational interactions that are established, at run-time, between business partners. This is why software development in the service-oriented paradigm requires new abstractions, methods and techniques.

The challenge that we face, and on which we wish to report, is to support this paradigm with mathematical foundations that allow us to define, in a rigorous and verifiable way, (1) the mechanisms through which modules can use externally procured services to offer services of their own, and (2) the way modules can be assembled into (sub-)systems that may, if desired, be offered themselves as (composite) modules. Having this goal in mind, we present in Section 2 a brief overview of the supported composition model and a summary of the different formal domains involved in it. Then, in Section 3, we formalise the notion of module as a graph labelled over the identified formal domains. Section 4 discusses the correctness property of modules and the notion of system as an assembly of modules. Finally, Section 5 develops the notion of composition through which composite modules are defined from systems.

2 The Composition Model

Modules in SRML-P are inspired by concepts proposed in Service Component Architectures (SCA) [10]. The main concern of SCA is to develop a middleware-independent architectural layer that can provide an open specification "allowing multiple vendors to implement support for SCA in their development tools and run-times". That is, SCA shares with us the goal of providing a uniform model of service behaviour that is independent of the languages and technologies used for programming and deploying services. However, whereas we focus on the mathematical structures that support this new architectural model, SCA looks "downstream" in the abstraction hierarchy and offers specific support for a variety of component implementation and interface types such as BPEL processes with WSDL interfaces, and Java classes with corresponding interfaces.

Given the complementarities of both approaches, we decided to stay as close as possible to the terminology and methodology of SCA. This is why in SRML-P we adopt the following formal domains when characterising the new architectural elements: business roles that type SCA *components*, business protocols that type SCA *external interfaces* (both entry points and required services), and interaction protocols that type SCA *internal wires*.

Service components do not provide any business logic: the units of business logic are *modules* that use such components to provide services when they are interconnected with a number of other parties offering a number of required services. In

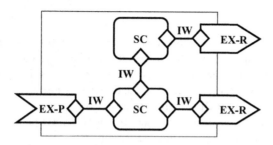

Fig. 1. A SRML-P module; SC–service component; EX-P–(provides) external interface; EX-R–(requires) external interface; IW–internal wire

a SRML-P module, both the provided services and those required from other parties are modelled as *external interfaces*, or interfaces for short. Each such interface specifies a stateful interaction (business protocol) between a service component and the corresponding party; that is, SRML-P supports both "syntactic" and "behavioural" interfaces.

The service components within a module orchestrate the interactions with the external parties that, in any given configuration, are linked to these interfaces. Like in SCA, modules are interconnected within systems by linking required external services of given modules with provided services offered by other modules. Such interconnections can be performed "just-in-time" once a mechanism is provided through which modules can be "discovered" and the binding of required with provided external interfaces can be effectively supported.

2.1 Business Roles

Central to SCA is the notion of *component*. In SRML-P, a component is a computational unit that fulfils a given *business role*, which is modelled in terms of an execution pattern involving a number of interactions that the component can maintain with other parties. We refer to the execution pattern as an *orchestration element*, or *orchestration* for short.

The model provided through the business role is independent of the language in which the component is programmed and the platform in which it is deployed; it may be a BPEL process, a Java program, a wrapped-up legacy system, *inter alia*. The orchestration is independent of the specific parties that are actually interconnected with the component in any given run-time configuration; a component is totally independent in the sense that it does not invoke services of any specific co-party – it just offers an interface of two-way interactions in which it can participate.

The primitives that we are adopting in SRML-P for describing business roles have been presented in [7] and, in more detail, also in [6]; they are defined in terms of typical event-condition-action rules in which the actions may involve interactions with other parties. An example is given in the Appendix in terms of a BookingAgent of a typical TravelBooking composite service. However, given that our focus in this paper is the notion of module, we do not need to commit to any specific orchestration language and, therefore, will not discuss the language used in SRML-P any further. All we need is to assume a set ***BROL*** of business roles to be given together with a number of mappings to other formal domains as detailed further on.

2.2 Signatures

One of the additional formal domains that we need to consider consists of the structures of interactions through which components can be connected to other architectural elements. These structures capture both classical notions of "syntactic" interface – i.e. declarations of types of interactions – and the ports through which interconnections are established. In SRML-P, interactions can be typed according to the fact that they are synchronous or asynchronous, and one or two-way; parameters can also be defined for the exchange of data during interactions.

We assume that such structures are organised in a category $SIGN$, the objects of which are called *signatures*. Morphisms of signatures define directional "part-of" relationships, i.e. a morphism $\sigma:S_1 \rightarrow S_2$ formalises the way a signature (structure of interactions) S_1 is part of S_2 up to a possible renaming of the interactions and corresponding parameters. In other words, a morphism captures the way the source is connected to the target, for instance how a port of a wire is connected to a component.

We assume that every business role defines a signature consisting of the interactions in which any component that fulfils the role can become involved. This is captured by a mapping $sign_{BROL}:BROL \rightarrow SIGN$. For instance, in the Appendix, we can see that the signature of a business role is clearly identified under "interactions". For simplicity, we do not give any detail of the categorical properties of signatures in SRML-P, which are quite straightforward.

We further assume that $SIGN$ is finitely co-complete. This means that we can compose signatures by computing colimits (amalgamated sums) of finite diagrams; typically, such diagrams are associated with the definition of complex structures of signatures, which can result from the way modules are put together as discussed in Section 0, or the way modules are interconnected as discussed in Section 5.

2.3 Business Protocols

Besides components, a module in SRML-P may declare a number of (external) interfaces. These provide abstractions (types) of parties that can be interconnected with the components declared in the module either to provide or request services; this is what, in SCA, corresponds to "Entry Points" and "External Services".

External interfaces are specified through *business protocols*, the set of which we denote by $BUSP$. Like business roles, protocols declare the interactions in which the external entities can be involved as parties; this is captured by a mapping $sign_{BUSP}:BUSP \rightarrow SIGN$. The difference with respect to business roles is that, instead of an orchestration, a business protocol provides a set of properties that model the protocol that the co-party is expected to adhere to. In the Appendix, we give as an example the business protocol that corresponds to the FlightAgent. Like for business roles, the signature of a business protocol in SRML-P is clearly identified under "interactions".

Business protocols, which model what in SCA corresponds to "external services", specify the conversations that the module expects relative to each party. Those that model what in SCA corresponds to an "entry point", specify constraints on the interactions that the module supports as a service provider. Examples of such constraints are the order in which the service expects invocations or deadlines for the user to commit, but also properties that the client may expect such as pledges on given parameters of the delivered service. It is the responsibility of the client to adhere to these protocols, meaning that the provider may not be ready to engage in interactions that are not according to the specified constraints.

2.4 Interaction Protocols

Service components and external interfaces are connected to each other within modules through *internal wires* that bind the interactions that both parties declare to support and coordinate them according to a given *interaction protocol*. Typically, an interaction protocol may include routing events and transforming data provided by a sender to the format expected by a receiver. The examples given in the Appendix are quite simple: they consist of straight synchronisations at the ports.

Just like business roles and protocols, an interaction protocol is specified in terms of a number of interactions. However, interaction protocols are somewhat more complex. On the one hand, an interaction protocol declares two disjoint sets of interactions; in SRML-P, this is done under the headings ROLE A and ROLE B. On the other hand, the properties of the protocol – what we called the *interaction glue* – are declared in a language defined over the union of the two roles, what we call its signature. We consider that we have a set *IGLU* of specifications of interaction glues together with a map $sign_{IGLU}:IGLU{\to}SIGN$.

In order to model the composition of modules, we also need a way of composing interaction protocols. For that purpose, we assume that *IGLU* is itself a co-complete category whose morphisms $\sigma:G_1{\to}G_2$ capture the way G_1 is a sub-protocol of G_2, again up to a possible renaming of the interactions and corresponding parameters. That is, σ identifies the glue that, within G_2, captures the way G_1 coordinates the

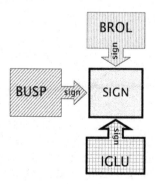

Fig. 2. How the different formal domains relate to each other: *BROL*–business roles; *BUSP*–business protocols; *IGLU*–interaction glue of protocols; *SIGN*–signatures

interactions $sign(G_1)$ as a part of $sign(G_2)$. More precisely, we assume that $sign_{IGLU}$ is a functor that preserves colimits, i.e. that the signature of a composition of protocols is a composition of their signatures.

2.5 Summary

The relationships between all these different formal domains are summarised in **Figure 2** (categories are represented with a thick line). For simplicity, we use *sign* as an abbreviated notation for $sign_{BROL}$, $sign_{BUSP}$ and $sign_{IGLU}$.

3 Defining Modules

As already mentioned, modules are the basic units of composition. They include external interfaces for required and provided services, and a number of components whose orchestrations ensure that the properties offered on the provides-interfaces are guaranteed by the connections established by the wires assuming that the services requested satisfy the properties declared on the requires-interfaces.

In our formal model, a module is defined as a graph: components and external interfaces are nodes of the graph and internal wires are edges that connect them. This graph is labelled by a function L: components are labelled with business roles, external interfaces with business protocols, and wires with connectors that include the specification of interaction protocols. An example of the syntax that we use in SRML-P for defining the graph and labelling function can be found in the Appendix.

Because a wire interconnects two nodes of the module (graph), we need some means of relating the interaction protocols used by the wire with the specifications (business roles or protocols) that label the nodes. The connection for a given node n and interaction protocol P is characterised by a morphism μ_n that connects one of the roles (A or B) of P and the signature $sign(L\ (n))$ associated with the node. We call a *connector* for a wire $n \xleftarrow{\ w\ } m$ a structure $<\mu_n, \pi_n, G, \pi_m, \mu_m>$ where G is the interaction glue of the protocol P and the morphisms π_n and π_m identify the roles of P:

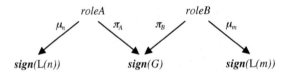

In SRML-P, connections are defined in a tabular form that should be self-explanatory as illustrated in the Appendix. Some wires may be labelled by more than one connector because they involve more than one interaction. In such cases, we may compose the connectors by taking the sum of their protocols. More concretely, if we have a collection $<\mu_n^i, \pi_n^i, G^i, \pi_m^i, \mu_m^i>$ of connectors labelling a wire $n \leftrightarrow m$, we can represent it by the connector $<\oplus\mu_n^i, \oplus\pi_n^i, \oplus P^i, \oplus\pi_m^i, \oplus\mu_m^i>$ given by the diagram:

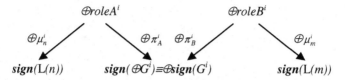

The morphisms are given uniquely by the properties of sums in *SIGN* [5]. This corresponds to looking at the set of connectors that labels a wire as defining a single connector, which makes it easier to define and manipulate modules.

Formally, we take a module *M* to consist of:

- A graph, i.e. a set *nodes(M)* and a set *wires(M)* of pairs $n \leftrightarrow m$ of nodes (elements of *nodes(M)*).
- A distinguished subset of nodes *requires(M)⊆nodes(M)*.
- At most one distinguished node *provides(M)∈nodes(M)\requires(M)*.
- A labelling function $\mathscr{L}$ such that:
  - $\mathscr{L}$ *(provides(M))∈BUSP* if *provides(M)* is defined
  - $\mathscr{L}$ *(n)∈BUSP* for every *n∈requires(M)*
  - $\mathscr{L}$ *(n)∈BROL* for every other node *n∈nodes(M)*
  - $\mathscr{L}$ $(n \leftrightarrow m)$ is a connector $<\mu_n, \pi_n, G, \pi_m, \mu_m>$.

A module *M* for which *provides(M)* is not defined corresponds to applications that do not offer any services but still require external services to fulfil their goals. They can be seen to be "agents" that, when bound to the external services that they require, execute autonomously in a given configuration as discussed below. Modules that do provide a service and can be discovered are called *service modules*. Notice that modules do not offer services to more than one user. However, multiple sessions may be allowed – an aspect that we do not address in this paper.

We can expand every wire $n \leftrightarrow m$ into the following labelled directed graph:

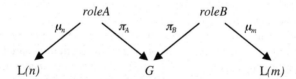

That is, we make explicit the protocol and the connections. We denote by *expanded(M)* the result of expanding all wires in this way. Therefore, in *expanded(M)* we have the nodes of *M* with the same labels – business roles and protocols – and, for each wire, an additional node labelled with a protocol, two additional nodes (ports) labelled with the roles of the protocol, and directed edges from the ports labelled with signature morphisms. For instance, the expanded graph of the module depicted in **Figure 1** has the following structure:

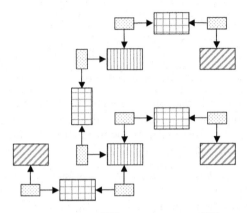

Fig. 3. The expanded graph of a module; ☐ – business role; ☐ – business protocol; ☐ – interaction glue; ☐ – signature (role)

4 Semantic Correctness

Section 3 defines some criteria that ensure the syntactic correctness of modules, namely the fact that the endpoints of the connectors in the wires match the labels of the nodes linked by the wire. In this section, we are concerned by the semantic correctness of service modules, i.e. the fact that the properties offered in the provides-interface are ensured by the orchestration of the components and the properties required of the other external interfaces.

The correctness condition is expressed in terms of logical entailment of properties of business protocols. The mechanisms that we provide for putting together, interconnecting and composing modules is largely independent of this logic. The particular choice of operators, their semantics and proof-theory are essential for supporting the modelling of service-based applications, i.e. for the pragmatics of "in-the-small" issues, but not for the semantics and pragmatics of modules as units of composition, i.e. for the "in-the-large" issues. What is important is that the logic satisfies some structural properties that are required for the correctness condition and the notion of module composition to work well together as explained below. In SRML-P, we use the temporal logic $\mu UCTL$ [8] defined over an alphabet of events such that every interaction declared in a signature gives rise to the following set of events (see [6] for additional explanations):

interaction🖰	The event of initiating *interaction*.
interaction⊠	The reply-event of *interaction*.
interaction✓	The commit-event of *interaction*.
interaction✗	The cancel-event of *interaction*.
interaction☀	The deadline-event of *interaction*.
interaction⚑	The revoke-event of *interaction*.

As a consequence, we assume that we have available an entailment system (or π-institution) [5,9] *<SIGN,gram, ⊢>* where ***gram:SIGN→SET*** is the grammar functor that, for every signature Q, generates the language used for describing properties of the interactions in Q. Notice that, given a signature morphism $\sigma:Q\rightarrow Q'$, $gram(\sigma)$ translates properties in the language of Q to the language of Q'. Notice that temporal logics define institutions [5].

We denote by $\vdash_Q$ the entailment system that allows us to reason about properties in the language of Q. We write $S\vdash_Q s$ to indicate that sentence s is entailed by the set of sentences S. Pairs $<Q,S>$ consisting of a set S of sentences over a signature Q – usually called theory presentations – can be organised in a category **SPEC** whose morphisms capture entailment. We denote by **sign** the forgetful functor that projects theories on the underlying signatures.

Given a specification $SP=<Q,S>$ and sets P and R_i of sentences over Q, we also use the notation

$$P\frac{\quad}{SP}\bigg|\begin{matrix}R_1\\\vdots\\R_N\end{matrix}$$

to indicate that $R_1\cup...\cup R_N\cup S\vdash_Q p$ for every $p\in P$, i.e. that the properties expressed by P are guaranteed by SP relying on the fact that the properties expressed in R_i hold.

As discussed in Section 2, the specifications of business roles, business protocols and interaction protocols carry a semantic meaning. We take this meaning to be defined by mappings **spec$_{BROL}$:BROL→SPEC**, **spec$_{BUSP}$:BUSP→SPEC** and **spec$_{IGLU}$: IGLU→SPEC** that, when composed with **sign:SPEC→SIGN**, give us the syntactic mappings discussed in Section 2.

In the case of business roles, this assumes that we can abstract properties from orchestrations, which corresponds to defining an axiomatic semantics of the orchestration language. In SRML-P, this means a straightforward translation of event-condition-action rules into $\mu UCTL$.

In the case of business and interaction protocols, this mapping is more of a translation from the language of external specifications to a logic in which one can reason about the properties of interactions as well as that of orchestrations. In SRML-P, the operators used in the examples given in the Appendix are translated as follows:

a **before** b	If b holds then a must have been true.	$AG(b \supset Pa)$
b **exceptif** a	b can occur iff b and a have never occurred.	$AG(\neg Pa\wedge H(\neg b) \equiv Eb)$
a **enables** b	b can occur iff a has already occurred but not b.	$AG(Pa\wedge H(\neg b) \equiv Eb)$
a **ensures** b	b will occur after a occurs, but b cannot occur without a having occurred.	$AG(b\supset Pa \wedge a\supset Fb)$

We further assume that the mapping **spec$_{IGLU}$** is in fact a functor, i.e. that the composition of interaction protocols preserves properties. This leads to the following extension of **Figure 2**:

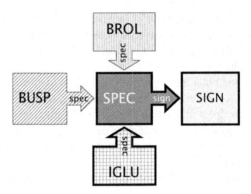

Fig. 4. Relating the specification domain with the other formal domains

The correctness property of service modules relies on the fact that the orchestrations of the business roles and the properties of the interaction protocols guarantee that the properties of the requires-interfaces entail those ensured by the provides-interfaces. To express it, we need a means of referring to the fragment of the module that is concerned with components and wires, what we call the *body* of the module. Formally, we define *body(M)* for a module *M* as being the diagram of specifications and signatures that is obtained from *expanded(M)* by applying the mappings ***spec*** to all the labels (business roles, business protocols and interaction protocols). That is, we obtain the same graph as that of *expanded(M)* except that we label the nodes with the specifications of the business roles and interaction protocols, and the signatures of the business protocols. For instance, the following picture corresponds to the body-diagram of the expanded-graph of **Figure 3:**

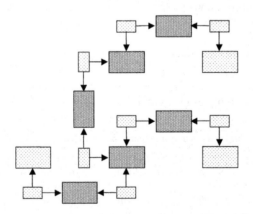

Fig. 5. The body diagram of a module

We assume that the category ***SPEC*** is finitely co-complete and coordinated over ***SIGN***, which allows us to calculate the colimit (amalgamated sum) of this diagram. The colimit returns a specification ***Body(M)*** and a morphism q_n:***sign***(L (n))→***sign***(M) for every node *n* of *expanded(M)*.

It is helpful to detail the construction of **Body(M)**. Its signature **sign(M)** is the colimit (amalgamation) of the diagram of signatures defined by *body(M)*. This signature contains all the interactions that are involved in the module; the morphisms q_n record in which nodes each interaction is used. The set of axioms of **Body(M)** consists of the union of the following sets:

- For every node n labelled by a business role W, the translation **gram**$(q_n)(S_W)$ where **spec**(W)=<**sign**$(W),S_W$>, i.e. we take the translations of the axioms of **spec**(W).
- For every node n labelled by a glue G of an interaction protocol, the translation **gram**$(q_n)(S_G)$ where **spec**(G)=<**sign**$(G),S_G$>, i.e. we take the translations of the axioms of **spec**(G).

Notice that the business protocols of the external interfaces are not used for calculating **Body(M)**: only their signatures are used. However, because their signatures are also involved, we can operate the same kind of translation on every external interface by using the corresponding signature morphism q:

- We denote by **Prov(M)** the translation of the specification of the business protocol of *provides(M)*, i.e. of the provide-interface of M.
- We denote by **Reqs**$_{1..N}$**(M)** the translations of the specifications of the business protocols in *requires(M)*, i.e. of the requires-interfaces of M.

Given that all these sets of sentences are now in the language of **sign(M)**, the correctness property of a service module M can be expressed by:

$$Prov(M) \frac{\qquad}{Body(M)} \left| \begin{array}{c} Reqs_1(M) \\ \vdots \\ Reqs_N(M) \end{array} \right.$$

That is, every property offered in the business protocol of the provides-interface must be entailed by the body of the module using the properties required in the business protocols of the requires-interfaces.

5   Composing Modules

In this section, we discuss the mechanisms through which modules can be assembled to create systems and modules can be created from systems. These mechanisms are similar to those provided in SCA, i.e. they provide a means of linking requires-external interfaces of a module with provides-external interfaces of other modules. In SRML-P, we provide only abstract models of such links, which we call *external wires*. That is, we remain independent of the technologies through which interfaces are bound to parties, which depend on the nature of the parties involved (BPEL processes, Java programs, databases, *inter alia*). In summary, external wires carry a proof-obligation to ensure that the properties offered by the provides-interface are implied by those declared in the requires-interfaces.

A system is a directed acyclic graph in which nodes are labelled by modules and edges are labelled with so-called "bindings" or "external wires". A binding for an edge $n \rightarrow k$ between modules M_n and M_k consists of:

- A node $r \in requires(M_n)$, i.e. one of the requires-interfaces of M_n. This node cannot be used by any other binding. Let this node be labelled with S_r.
- A specification morphism $\rho{:}spec(S_r) \to spec(S_p)$ where S_p is the business proto-col of $provides(M_k)$, i.e. of the provides-interface of M_k.

In other words, bindings connect a requires-interface of one module to the pro-vides-interface of another module such that the properties of the requires-interface are implied by the properties of the provides-interface.

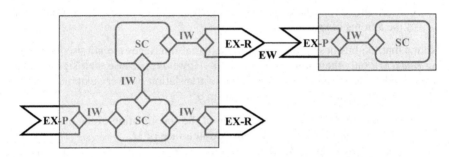

Fig. 6. An assembly of modules defining a SRML-P system; EW–external wire

SRML-P also supports a way of offering a system as a module, i.e. of turning an assembly of services into a composite service that can be published and discovered on its own. The operation that collapses a system into a module internalises the external wires and forgets the external specifications.

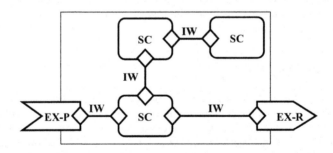

Fig. 7. The previous system turned into a module

Formally, a module may be created from every (finite) weakly connected system by internalising the bindings. The resulting module M is as follows:

- The graph of M is obtained from the sum (disjoint union) of the graphs of all modules involved in the system by eliminating, for every edge $n \to k$ of the sys-tem, the nodes r (requires) of M_n and $provides(M_k)$, and adding, for every such edge $n \to k$ of the system, an edge $i \leftrightarrow j$ between any two nodes i and j such that $i \leftrightarrow r$ is an edge of M_n and $provides(M_k) \leftrightarrow j$ is an edge of M_k.

- The labels are inherited from the graphs of the modules involved, except for the new edges $i \leftrightarrow j$. These are calculated by merging the connectors that label $i \leftrightarrow r$ and $provides(M_k) \leftrightarrow j$. The interaction protocol of the new connector is obtained through the colimit diagram below where $m = provides(M_k)$.

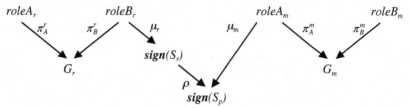

This composition is defined by the following colimit diagram in **IGLU**:

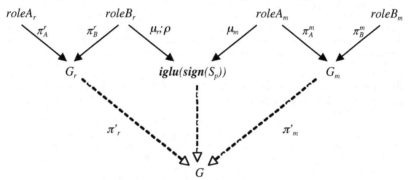

The rest of the connector is defined by the morphisms μ_i of $i \leftrightarrow r$ and μ_j of $provides(M_k) \leftrightarrow j$:

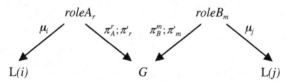

- *requires(M)* consists of the remaining requires-interfaces.
- *provides(M)* consists of the remaining provides-interface, if one remains. Notice that the connectivity of the graph implies that at most one provides-interface can remain.

The colimits calculated in order to obtain the protocol of the new connectors are expressed over a "diagram" that involves both signatures (those of the external interfaces and the ports) and protocols.

For this construction to make sense, we assume that the category **IGLU** is coordinated over **SIGN** [5]. This means that we have a canonical way of lifting signatures to interaction protocols that respects the interactions. In other words, every signature can be regarded as an interaction protocol through which the "diagram" above defines a diagram in **IGLU**, thus allowing for the colimit to be computed.

The following picture depicts the graph involved in the composition considered in **Figure 6:**

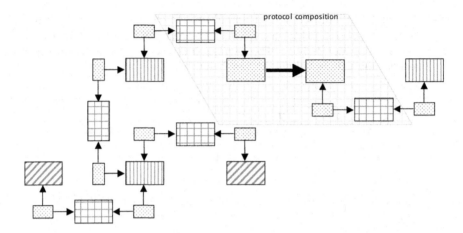

Fig. 8. The graphs involved in a composition; the diagram of interaction protocols involved in the internalisation of the binding is singled out

The graph obtained from the internalisation of the binding is the one that expands the module identified in **Figure 7:**

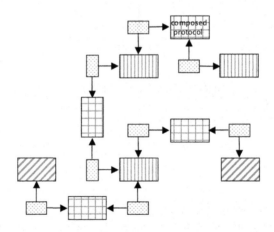

Fig. 9. The expanded graph of the composite module

6   Concluding Remarks and Further Work

In this paper, we have described some of the primitives that are being proposed for the SENSORIA Reference Modelling Language in order to support building systems in service-oriented architectures using "technology agnostic" terms. More specifically, we have focused on the language that supports the underlying composition model. This is a minimalist language that follows a recent proposal for a Service Component Architecture [10] that "builds on emerging best practices of removing or abstracting middleware programming model dependencies from business logic". However, whereas the SCA-consortium concentrates on the definition of an open specification that supports a variety of component implementation and interface types, and on the deployment, administration and configuration of SCA-based applications, our goal is to develop a mathematical framework in which service-modelling primitives can be formally defined and application models can be reasoned about.

Our composition model relies on the notion of module, which we adapted from SCA. Modules can be discovered and bound to other modules at run-time to produce configurations. We proposed a formal model for module assembly and composition in line with algebraic notions of component such as [2] and [4]. The former proposes a notion of component that is similar to what is put forward by SCA but misses the notion of module as providing services that result from the orchestration of components and external services. Its algebraic semantics is based on Interface Automata [1], which are similar to I/O-automata, defined over operation (method) invocations; as explained in Section 4 (see also [6]), SRML works over a richer alphabet of events that capture the kind of stateful interactions typical of services. The latter [4] is based on more traditional algebraic notions of module [3] and uses graph-based formalisms to model component behaviour. The underlying algebraic framework is, once again, similar to the one we use but some research effort needs to be dedicated to bring out similarities and complementarities.

We are currently developing a notion of configuration for SRML-P as a collection of components wired together that models a run-time composition of service components. A configuration results from having one or more clients using the services provided by a given module, possibly resulting from a complex system, with no external interfaces, i.e. with all required external interfaces wired-in. It is at the level of configurations that we address run-time aspects of service composition such as service discovery (and service-level agreements), sessions (and dynamic reconfiguration), as well as notions of persistence.

Acknowledgments

J. Fiadeiro was partially supported by a grant from the Royal Society (UK) while on study leave from the University of Leicester at the University of Pisa during April and May 2006. A. Lopes was partially supported by the Foundation for Science and Technology (FCT, Portugal) during an extended stay at the University of Pisa during May 2006. We wish to thank the referees for many useful remarks, especially for pointing us to [2].

References

1. L. de Alfaro, T. Henzinger (2001) Interface automata. *ESEC/SIGSOFT_FSE*. ACM Press, New York, pp 109–120
2. H. Baumeister, F. Hacklinger, R. Hennicker, A. Knapp, M. Wirsing (2005) A component model for architectural programming. *Electronic Notes in Theoretical Computer Science* to appear
3. H. Ehrig, B. Mahr (1990) *Fundamentals of Algebraic Specification 2: Module Specifications and Constraints*. EATCS Monographs on Theoretical Computer Science, vol 21. Springer, Berlin Heidelberg New York
4. H. Ehrig, F. Orejas, B. Braatz, M. Klein, M. Piirainen (2004) A component framework for system modeling based on high-level replacement systems. *Software Systems Modeling* 3:114–135

5. J. L. Fiadeiro (2004) *Categories for Software Engineering.* Springer, Berlin Heidelberg New York

6. J. L. Fiadeiro, A. Lopes, L. Bocchi (2006) *The SENSORIA Reference Modelling Language: Primitives for Service Description and Composition.* Available from www.sensoria-ist.eu

7. J. L. Fiadeiro, A. Lopes, L. Bocchi (2006) A formal approach to service-oriented architecture. In: M. Bravetti, M. Nunez, G. Zavattaro (eds) *Web Services and Formal Methods. LNCS, vol 4184.* Springer, Berlin Heidelberg New York, pp 193–213

8. S. Gnesi, F. Mazzanti (2005) A model checking verification environment for UML State-charts. In: *Proceedings of XLIII Congresso Annuale AICA "Comunita' Virtuale dalla Ricerca all'Impresa dalla Formazione al Cittadino",* University of Udine – AICA (paper available from fmt.isti.cnr.it)

9. J. Goguen, R. Burstall (1992) Institutions: abstract model theory for specification and programming. *Journal ACM* 39(1):95–146

10. SCA Consortium (2005) *Building Systems using a Service Oriented Architecture.* White-paper available from www-128.ibm.com/developerworks/library/specification/ws-sca (version 0.9)

Appendix – TravelBooking

In this Appendix, we provide parts of a typical travel-booking process involving a flight and a hotel agent. The module – *TRAVELBOOKING* – that defines this composite service exposes to the environment an interface for booking a flight and a hotel for a given itinerary and dates. External services are requested in order to offer the service behaviour that the module declares to provide.

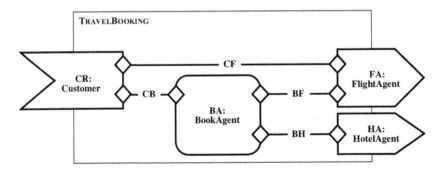

TRAVELBOOKING consists of:

- CR – the external interface of the service provided by the module, of type *Customer*;
- FA – the external interface of a service required for handling the booking of flights, of type *FlightAgent*;
- HA – the external interface of a service required for handling the booking of hotels, of type *HotelAgent*;

- BA – a component that coordinates the business process, of type *BookAgent*;
- *CB, CF, BF, BH* – four internal wires that make explicit the partner relationship between *CR* and *BA*, *CR* and *FA*, *BA* and *FA*, and *BA* and *FA*.

MODULE TravelBooking **is**

COMPONENTS

   BA: BookAgent

PROVIDES

   CR: Customer

REQUIRES

   FA: FlightAgent
   HA: HotelAgent

WIRES

BA BookAgent		**BF**		**FA** FlightAgent	
s&r bookFlight	S_1		R_1	**r&s** lockFlight	
⌂ from	i_1	**Straight**	i_1	⌂ from	
to	i_2	I[airport,airport,	i_2	to	
out	i_3	date,date]	i_3	out	
in	i_4	O[fref]	i_4	in	
✉ fconf	o_1		o_1	✉ fconf	
rcv fConfirm	R_1	**Straight**	S_1	**snd** flightAck	
⌂ result	i_1	I[bool]	i_1	⌂ result	

[...]

SPECIFICATIONS

BUSINESS ROLE BookAgent **is**

INTERACTIONS
> **r&s** bookTrip
> > ⌂ from,to:airport, out,in:date
> > ⊠ tconf:(fcode,hcode)
>
> **s&r** bookFlight
> > ⌂ from,to:airport, out,in:date
> > ⊠ fconf:fcode
>
> **s&r** bookHotel
> > ⌂ checkin:date, checkout:date
> > ⊠ hconf:hcode
>
> **snd** tAck
> > ⌂ result:bool
>
> **rcv** fConfirm
> > ⌂ result:bool

State variables for storing data that may be needed during the orchestration.

s is used for control flow, i.e. for encoding an internal state machine.

The other state variables are used for storing data transmitted through the parameters of interactions.

ORCHESTRATION
> **local** s:[0..6], fconf:fcode, hconf:hcode,
> out,in:date,   from,to:airport
> frep, hrep: boolean
>
> **initialisation** s=0
>
> **termination** s=3 ∨ (s=6 ∧ today≥out)
>
> **transition** TOrder
> > **triggeredBy** bookTrip⌂?
> > **guardedBy** s=0
> > **effects** from'=bookTrip.from
> > > ∧ to'=bookTrip.to
> > > ∧ out'=bookTrip.out
> > > ∧ in'=booktrip.in
> > > ∧ out'≥today ⊃ s'=1
> > > ∧ out'<today ⊃ s'=3
> >
> > **sends** out'>today ⊃ bookFlight⌂!
> > > ∧ bookFlight.from=from'
> > > ∧ bookFlight.to=to'
> > > ∧ bookFlight.out=out'
> > > ∧ bookflight.in=in'
> > > ∧ alertDate⌂!
> > > ∧ alertDate.Ref="flight"
> > > ∧ alertDate.Interval=fresp
> > > ∧ out'≤today ⊃ bookTrip⊠!
> > > ∧ bookTrip.Reply=false

Property guaranteed for the initial state.

Property that determines when the orchestration has terminated.

If *var* is a state variable, *var'* denotes its value after the transition; this expression can be used in both "effects" and "sends".

A request to travel on a date already passed leads immediately to a final state.

today is an external service that we assume to be globally available; it provides the current date.

In response to a request for travelling in a future date, a flight request is issued and a timeout is set with the duration that the agent is willing to wait for a reply.

alertDate is also a service that is globally available; it replies when the duration set-up in the parameter *Interval* elapses. We use the parameter *Ref* to correlate different alerts that are sent.

[...]

BUSINESS PROTOCOL FlightAgent **is**

   INTERACTIONS

     r&s lockFlight
       ⌂ from,to:airport, out,in:date
       ⊠ fconf:fcode

     s&r payRequest
       ⌂ amount:nat, benef:account,
        bank:servRef
       ⊠ pay:payData

     snd payAck
       ⌂ result:bool

     snd payRefund
       ⌂ amount:nat

   BEHAVIOUR (fref:string)
     lockFlight⌂? **exceptif** true
     lockFlight⊠! ∧ lockFlight.Reply
       ⊃ alertDate⌂!
         ∧ alertDate.Ref=fref
         ∧ alertDate.Interval≥fval
     lockFlight●⃰! ⊃ alertDate⊠?
       ∧ alertDate.Ref=fref
     lockFlight✓? **ensures** payRequest⌂!
     payRequest⊠? ∧ payRequest.Reply
                  ensures payAck!
     today<lockFlight.out ⊃ (payAck!
       ∧ payAck.result **enables** lockFlight⊹?)
     lockFlight⊹? **ensures** payRefund⌂!
       ∧ payRefund.amount
             >payRequest.amount*0.9

Side notes:

In the initial state, FA is required to be ready to receive a request for a flight.

The timeout for flight reservations is at least fval.

FA is required to request the payment after receiving the commit.

FA is required to send *payAck* to acknowledge the reception of a successful payment.

FA is required to accept the revoke of a flight booking until the day of departure and provide a refund of at least 90% of its cost.

INTERACTION PROTOCOL Straight.I(d_1,d_2,d_3,d_4)O(d_5) **is**

   ROLE A
     s&r S_1
       ⌂ i_1:d_1, i_2:d_2, i_3:d_3, i_4:d_4
       ⊠ o_1:d_4

   ROLE B
     r&s R_1
       ⌂ i_1:d_1, i_2:d_2, i_3:d_3, i_4:d_4
       ⊠ o_1:d_4

   COORDINATION
     $S_1 \equiv R_1$
     $S_1.i_1=R_1.i_1$
     $S_1.i_2=R_1.i_2$
     $S_1.i_3=R_1.i_3$
     $S_1.i_4=R_1.i_4$
     $S_1.o_1=R_1.o$

Autonomous Units and Their Semantics - The Parallel Case*

Hans-Jörg Kreowski and Sabine Kuske

University of Bremen, Department of Computer Science
P.O. Box 330440, D-28334 Bremen, Germany
{kreo,kuske}@informatik.uni-bremen.de

Abstract. Communities of autonomous units are rule-based and graph-transformational devices to model data-processing systems that may consist of distributed and mobile components. The components may communicate and interact with each other, they may link up to ad-hoc networks. In this paper, we introduce and investigate the parallel-process semantics of communities of autonomous units.

1 Introduction

Communities of autonomous units are introduced in [1] as rule-based and graph-transformational devices to model processes that run interactively, but independently of each other in a common environment. The main goal of this approach is to cover new programming and modeling paradigms like communication networks, multiagent systems, swarm intelligence, ubiquitous, wearable and mobile computing in a common and systematic way with rigorous formal semantics. While the sequential process semantics is considered in [1], we introduce and start to investigate the parallel-process semantics of communities of autonomous units in this paper.

An autonomous unit consists of a goal, a set of rules, and a control condition. The rules can be applied to environments which are assumed to be graphs. Rule application is usually quite nondeterministic because many rules may be applicable to an environment and even a single rule may be applicable to various parts of the environment. The control condition can cut down this nondeterminism by dividing all possible rule applications into the "good" and the "bad" ones. The control condition gives the unit autonomy in the sense that the unit can decide for one of the good rule applications to be performed. The goal describes the environments the unit wants to reach. A set of autonomous units forms a community which is additionally provided with a description of initial environments (where computational processes can start) and with an overall goal. A process then consists of a finite or infinite sequence of rule applications.

* Research partially supported by the Collaborative Research Centre 637 (Autonomous Cooperating Logistic Processes: A Paradigm Shift and Its Limitations) funded by the German Research Foundation (DFG).

J.L. Fiadeiro and P.-Y. Schobbens (Eds.): WADT 2006, LNCS 4409, pp. 56–73, 2007.

To cover parallelism, we assume that not only single rules but multisets of rules can be applied to environment graphs. This means that in each step many rules can be applied and single rules multiple times. As the rules may belong to different units, the autonomous units act in parallel. The units in a community are not directly aware of each other, but they may notice the outcome of the activities of their co-units because some of their rules may become applicable and others may loose this possibility. In this way, autonomous units can communicate and interact. To cover these phenomena in the process semantics of a single unit, we assume a change relation on environments that makes the environment dynamic. Then a parallel process of a single autonomous unit can be described as a sequence of application of multisets of rules of this unit in parallel with changes of the environment. These notions are introduced in Sections 3 and 4 and illustrated by the running example of a community of two autonomous units that work together and compute shortest paths. The basic ingredients including graphs, rules, rule applications, graphs class expressions to describe goals, and control conditions are defined by means of the notion of graph transformation approaches in Section 2. All components are quite generic so that they can be instantiated in various ways according to need or taste. Graph transformation approaches play a somewhat similar role for graph transformation than institutions for algebraic specification.

To shed some first light on the significance and usefulness of communities of autonomous units with parallel-process semantics, we compare our concepts with the parallelism provided by other well-known frameworks. In Section 5, we translate place/transition systems into communities of autonomous units and show that firing sequences of multisets of transitions correspond to parallel processes of the associated community. Similarly, cellular automata can be considered as communities of autonomous units as shown in Section 6. Cellular automata are particularly interesting as all their cells change states simultaneously so that the mode of computation is massively parallel. In Section 7, we discuss the relationship between communities of autonomous units and multiagent systems. As the latter are defined in an axiomatic way, the former can be seen as rule-based models providing an operational semantics for multiagent systems independent of the implementation of agents.

The introduction and investigation of autonomous units is mainly motivated by the Collaborative Research Centre 637 *Autonomous Cooperating Logistic Processes*. This interdisciplinary project focuses on the question whether logistic processes with autonomous control may be more advantageous than those with central control, especially regarding time, costs and robustness. The guiding principle of autonomous units is the integration of autonomous control into rule-based models of processes. The aims are

1. to describe algorithmic and particularly logistic processes in a very general and uniform way, based on a well-founded semantic framework,
2. to provide a range of applications that reaches from classical process chain models like the ones by Kuhn (see, e.g., [2]) or Scheer (see, e.g., [3]) and

the well-known Petri nets (see, e.g., [4,5]) to agent systems see, e.g., [6]) and swarm intelligence (see, e.g., [7]),
3. to comprise the foundation of the dynamics of processes by means of rules where rule applications define process, transformation, and computation steps yielding local changes.

Archetypes of a rule-based approach to data processing are grammatical systems of all kinds (see, e.g., [8]) and term rewriting systems (see, e.g., [9]) as well as the domain of graph transformation (see, e.g., [10,11,12]) and DNA computing (see, e.g., [13]). The rule-based approach is meant to ensure an operational semantics as well as to lay the foundation for formal verification.

In [1] we have shown that autonomous units with sequential process semantics generalize our former modeling concept of graph transformation units (see, e.g., [14]). While the latter apply their rules without any interference from the outside, an autonomous unit works in a dynamic environment which may change because of the activities of other units in the community. This makes quite a difference because the running of the system is no longer controlled by a central entity. Clearly, this applies to the parallel case, too, because it generalizes the sequential case.

2 Parallel Graph Transformation Approaches

Graph transformation (see e.g. [10,15]) constitutes a formal specification technique that supports the modeling of the rule-based transformation of graph-like, diagrammatic, and visual structures in an intuitive and direct way. The ingredients of graph transformation are provided by so-called graph transformation approaches. In this section, we recall the notion of a graph transformation approach as introduced in [14] but modified with respect to the purposes of this paper.

Two basic components of every graph transformation approach are a class of graphs and a class of rules that can be applied to these graphs. In many cases, rule application is highly nondeterministic – a property that is not always desirable. Hence, graph transformation approaches can also provide a class of control conditions so that the degree of nondeterminism of rule application can be reduced. Moreover, graph class expressions can be used in order to specify for example sets of initial and terminal graphs of graph transformation processes.

The basic idea of parallelism in a rule-based framework is the application of many rules simultaneously and also the multiple application of a single rule. To achieve these possibilities, we assume that multisets of rules can be applied to graphs rather than single rules.

Given some basic domain D, the set of all multisets D_* over D with finite carriers consists of all mappings $m: D \to \mathbb{N}$ such that the carrier $car(m) = \{d \in D \mid m(d) \neq 0\}$ is finite. For $d \in D, m(d)$ is called the multiplicity of d in m. The union or sum of multisets can be defined by adding corresponding multiplicities. D_* with this sum is the free commutative monoid over D where the multiset with empty carrier is the null element, i.e. $null\colon D \to \mathbb{N}$ with $null(D) = 0$. Note that the elements of D correspond to singleton multisets, i.e. for $d \in D, \hat{d}\colon D \to \mathbb{N}$ with $\hat{d}(d) = 1$ and $\hat{d}(d') = 0$ for $d' \neq d$. If $\mathcal{R}$ is a set of rules, $r \in R_*$ comprises a selection of rules each with some multiplicity. Therefore, an application of r to a graph yielding a graph models the parallel and multiple application of several rules.

Formally, a parallel graph transformation approach is a system consisting of the following components.

- $\mathcal{G}$ is a class of *graphs*.
- $\mathcal{R}$ is a class of *graph transformation rules* such that every $r \in R_*$ specifies a binary relation on graphs $SEM(r) \subseteq \mathcal{G} \times \mathcal{G}$.
- $\mathcal{X}$ is a class of *graph class expressions* such that each $x \in \mathcal{X}$ specifies a set of graphs $SEM(x) \subseteq \mathcal{G}$.
- $\mathcal{C}$ is a class of *control conditions* such that each $c \in \mathcal{C}$ specifies a set of sequences $SEM_{Change}(c) \subseteq SEQ(\mathcal{G})$ where $Change \subseteq \mathcal{G} \times \mathcal{G}$.[1] As we will see later the relation $Change$ defines the changes that can occur in the environment of an autonomous unit. Hence, control conditions have a loose semantics which depends on the changes of the environment given by $Change$.

For technical simplicity we assume in the following that $\mathcal{A} = (\mathcal{G}, \mathcal{R}, \mathcal{X}, \mathcal{C})$ is an arbitrary but fixed parallel graph transformation approach. The multisets of rules in R_* are called parallel rules. A pair of graphs $(G, G') \in SEM(r)$ for some $r \in R_*$ is an application of the parallel rule r to G with the result G'. It may be also called a direct parallel derivation or a parallel derivation step.

Sometimes it is meaningful to parameterize the semantics of a graph class expression x by the class of graphs. i.e. $SEM_G(x) \subseteq \mathcal{G}$ for all $G \in \mathcal{G}$. This allows one to describe relations and functions between graphs rather than just sets of graphs. An example of this kind can be found in Section 4.

Examples

In the following we present some instances of the components of parallel graph transformation approaches. These will be used in the following sections. Further examples of graph transformation approaches can be found in e.g. [10].

Graphs. A well-known instance for the class $\mathcal{G}$ is the class of all directed edge-labeled graphs. Such a graph is a system $G = (V, E, s, t, l)$ where V is a set of nodes, E is a set of edges, $s, t\colon E \to V$ assign to every edge its source $s(e)$ and its target $t(e)$, and the mapping l assigns a label to every edge in E. The components of G are also denoted by V_G, E_G, etc. As usual, a graph M is a subgraph of G, denoted by $M \subseteq G$ if $V_M \subseteq V_G$, $E_M \subseteq E_G$, and s_M, t_M, and l_M are the restrictions of s_G, t_G, and l_G to E_M. A graph morphism $g\colon L \to G$ from a graph L to a graph G consists of two mappings $g_V\colon V_L \to V_G$, $g_E\colon E_L \to E_G$ such that sources, targets and labels are preserved, i.e. for all $e \in E_L$, $g_V(s_L(e)) = s_G(g_E(e))$,

---

[1] For a set A 2^A denotes its powerset and $SEQ(A)$ the set of finite and infinite sequences over A.

$g_V(t_L(e)) = t_G(g_E(e))$, and $l_G(g_E(e)) = l_L(e)$. In the following we omit the subscript V or E of g if it can be derived from the context.

Other classes of graphs are trees, forests, Petri nets, undirected graphs, hypergraphs, etc.

Rules. As a concrete example of rules we consider the so-called dpo-rules each of which consists of a triple $r = (L, K, R)$ of graphs such that $L \supseteq K \subseteq R$. The application of a rule to a graph G yields a graph G', if one proceeds according to the following steps: (1) Choose a graph morphism $g: L \to G$ so that for all items x, y (nodes or edges) of L $g(x) = g(y)$ implies that x and y are in K. (2) Delete all items of $g(L) - g(K)$ provided that this does not produce dangling edges. (In the case of dangling edges the morphism g cannot be used.) (3) Add R to the resulting graph D, and (4) glue D and R by identifying the nodes and edges of K in R with their images under g. The conditions of (1) and (2) concerning g are called gluing condition.

Graph transformation rules can be depicted in several forms. In the following they are either shown in the form $L \supseteq K \subseteq R$ or by drawing only its left-hand side L and its right-hand side R together with an arrow pointing from L to R, i.e. $L \to R$. The different nodes of K are distinguished by different fill-styles.

A graph transformation rule (L, K, R) with positive context is a quadruple (PC, L, K, R) such that $L \subseteq PC$. It can be applied to G by applying (L, K, R) to G as described provided that there is a morphism $g': PC \to G$ such that the restriction of g' to L equals g. In the folllowing, a rule with positive context is depicted as $PC \supseteq L \supseteq K \subseteq R$ where different fill-styles determine the nodes and edges of L in PC. A graph transformation rule with negative context is defined as (NC, L, K, R) where (L, K, R) is a rule and $L \subseteq NC$. It can only be applied to G if the negative context of L is not in G, i.e. if the morphism $g: L \to G$ cannot be extended to some morphism $g': NC \to G$ of which g is the restriction to L (cf. also [16]). Rules with negative context are depicted as $NC \to R$ such that the part of NC not belonging to L is dashed (see for example Fig. 1).

Given two rules $r_i = (L_i, K_i, R_i)$ $(i = 1, 2)$ their parallel composition yields the rule $r_1 + r_2 = (L_1 + L_2, K_1 + K_2, R_1 + R_2)$ where $+$ denotes the disjoint union of graphs. In the same way one can construct a parallel rule from any multiset $r \in \mathcal{R}_*$. For every pair $(G, G') \in SEM(r_1 + r_2)$ there exist graphs M_1 and M_2 such that (G, M_1) and (M_2, G') are in $SEM(r_1)$ and (G, M_2) and (M_1, G') are in $SEM(r_2)$. This means that the graph G' can also be obtained from G by applying the rules r_1 and r_2 sequentially and in any order. Moreover, let r_i $(i = 1, 2)$ be two (parallel) rules and let $g_i: L_i \to G$ be two morphisms that satisfy the gluing condition described in steps (1) and (2) of a rule application. Then r_1 and r_2 are independent w.r.t. g_i if the the following independence condition is satisfied:

$$g_1(L_1) \cap g_2(L_2) \subseteq g_1(K_1) \cap g_2(K_2).$$

In this case both rules can be applied to G in parallel via the application of $r_1 + r_2$ using the graph morphism $\langle g_1 + g_2 \rangle: L_1 + L_2 \to G$ such that $\langle g_1 + g_2 \rangle(x) = g_i(x)$ if x is an element of L_i (see, e.g., [17] for more details).

Graph class expressions. Every subset $M \subseteq \mathcal{G}$ is a graph class expression that specifies itself, i.e. $SEM(M) = M$. Moreover, every set $\mathcal{L}$ of labels specifies the class of all graphs in $\mathcal{G}$ the labels of which are elements of $\mathcal{L}$. Every set $P \subseteq \mathcal{R}_*$ of (parallel) graph transformation rules can also be used as a graph class expression specifying the set of all graphs that are reduced w.r.t. P where a graph is said to be reduced w.r.t. P if no rules of P can be applied to the graph. The least restrictive graph class expression is the term *all* specifying the class $\mathcal{G}$.

Control conditions. The least restrictive control condition is the term *free* that allows all parallel graph transformations, i.e. $SEM_{Change}(free) = SEQ(\mathcal{G})$ for all $Change \subseteq \mathcal{G} \times \mathcal{G}$. Another useful control condition is $alap(P)$ where $P \subseteq \mathcal{R}_*$. It applies P as long as possible. More precisely, for every $Change \subseteq \mathcal{G} \times \mathcal{G}$ $SEM_{Change}(alap(P))$ consists of all finite sequences $(G_0, \ldots G_n) \in SEQ(\mathcal{G})$ for which there is an $i \in \{0, \ldots, n\}$ such that no rule in P can be applied to the graphs in $(G_i, \ldots, G_n)$. The condition $alap(P)$ can also be used to specify infinite sequences, a more complicated case that is not needed here.

3 Autonomous Units

Autonomous units act within or interact on a common environment which is modeled as a graph. An autonomous unit consists of a set of graph transformation rules, a control condition, and a goal. The graph transformation rules contained in an autonomous unit *aut* specify all transformations the unit *aut* can perform. Such a transformation comprises for example a movement of the autonomous unit within the current environment, the exchange of information with other units via the environment, or local changes of the environment. The control condition regulates the application process. For example, it may require that a sequence of rules be applied as long as possible or infinitely often. The goal of a unit is a graph class expresson determining how the transformed graphs should look like.

Definition 1 (Autonomous unit). An *autonomous unit* is a system $aut = (g, P, c)$ where $g \in \mathcal{X}$ is the *goal*, $P \subseteq \mathcal{R}$ is a set of graph transformation rules, and $c \in \mathcal{C}$ is a control condition. The components of *aut* are also denoted by g_{aut}, P_{aut}, and c_{aut}, respectively.

An autonomous unit modifies an underlying environment while striving for its goal. Its semantics consists of a set of transformation processes being finite or infinite sequences of environment transformations. An environment transformation comprises the parallel application of local rules or environment changes typically performed by other autonomous units that are working in the same environment. These environment changes are given as a binary relation of environments. Because the parallel-process semantics is meant to describe the simultaneous activities of autonomous units, the environment changes must be possible while a single autonomous unit applies its rules. To achieve this, we assume that there are some rules, called metarules, the application of which defines environment changes. Consequently, environment changes and ordinary rules can be applied in parallel. Hence, in this parallel approach a transformation process of an autonomous unit consists of a sequence of parallel rule applications which combine local rule applications with environment changes specified by other components. Every autonomous unit has exactly one thread of control. Autonomous units regulate their transformation processes by choosing in every step only those rules that are allowed by its control condition. A finite transformation process is called successful if its last environment satisfies the unit goal. Every infinite transformation process is successful if it contains infinitely many environments that satisfy the goal.

Definition 2 (Parallel semantics)

1. Let $aut = (g, P, c)$ be an autonomous unit and let $Change \subseteq \mathcal{G} \times \mathcal{G}$. Let $\mathcal{MR} \subseteq \mathcal{R}_*$ be a set of parallel rules, called metarules, such that $SEM(\mathcal{MR}) = \bigcup\limits_{r \in \mathcal{MR}} SEM(r) = Change$. Let $s = (G_0, G_1, G_2, \cdots) \in SEQ(\mathcal{G})$.
   Then $s \in PAR_{Change}(aut)$ if
   - for $i = 0, \cdots, |s|$ if s is finite[2] and for $i \in \mathbb{N}$ if s is infinite, $(G_{i-1}, G_i) \in SEM(r + r')$ for some $r \in P_*$ and $r' \in \mathcal{MR}$,
   - $s \in SEM_{Change}(c)$.
2. The sequence s is called a *successful transformation process* if s is finite and $G_{|s|} \in SEM(g)$ or there is an infinite monotone sequence $i_0 < i_1 < i_2 < \cdots$ with $G_{i_j} \in SEM(g)$ for all $j \in \mathbb{N}$.

The elements of $PAR_{Change}(aut)$ are sequences of applications of parallel rules which may be called the parallel processes of aut. Every single step of these processes applies a parallel rule of the form $r + r'$ where r is a parallel rule of the unit aut and r' is a metarule. Therefore, while the autonomous unit acts on the environment graph, the environment may change in addition. But as r and r' may be the null rule and $r + null = r$ as well as $null + r' = r'$, a step can also be an exclusive activity of aut or a change of the environment only.

Examples
As examples of autonomous units consider the units *minimum* and *sum* depicted in Fig. 1. The underlying graphs are labeled with natural numbers representing distances. The graph class expression for the goals of both units is *all* meaning that both units do not have any particular goal. The rule of *minimum* deletes the longer one out of two parallel edges labeled with natural numbers. The control condition of the unit *minimum* requires that the rule be applied as long as possible. In other words, a *minimum* process can only stop if no parallel edges are around. Two rule applications are independent if they delete different edges. This means that the rule can be applied k times in parallel, if the corresponding

---

[2] For a finite sequence s its number of elements is denoted by $|s|$.

parallel rule application deletes k edges. In particular, one can transform each graph in a simple one without parallel edges in a single step.

The rule of the second unit *sum* can be applied to a path e_1, e_2 provided that e_1 and e_2 are labeled with natural numbers x and y and that there exists no edge from the source of e_1 to the target of e_2 that is labeled with a number $z \leq x+y$. The last requirement is expressed by the dashed edge which represents negative context. The rule inserts a new edge from the source of e_1 to the target of e_2 and labels it with $x + y$. This rule must also be applied as long as possible. Moreover, the rule can only be applied if the graph morphism from the left-hand side of the rule to the current graph is injective. This means that the rule can be applied neither to edges that are loops nor to a cycle of length two. This condition ensures that no loops are produced in the computation of the sums of the edge labels. Each two applications of the *sum*-rule are independent because nothing is deleted. Consequently, the *sum*-rule can be applied k times in parallel for every $k \in \mathbb{N}$ as long as there are loop-free paths of length 2 satisfying the negative application condition. In particular, a parallel rule can be applied so that afterwards each loop-free path of length 2 and distance $x + y$ has got a parallel edge of distance $z \leq x + y$.

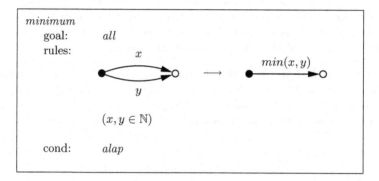

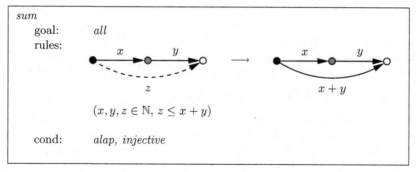

Fig. 1. Two autonomous units

4 Communities of Autonomous Units

Autonomous units are meant to work within a community of autonomous units that modify the common environment together. In the parallel case these modifications take place in an interleaving manner. Every community is composed of an overall goal that should be achieved, an environment specification that specifies the set of initial environments the community may start working with, and a set of autonomous units. The overall goal may be closely related to the goals of the autonomous units in the community. Typical examples are the goals admitting only graphs that satisfy the goals of one or all autonomous units in the community.

Definition 3 (Community). A *community COM* is a triple $(Goal, Init, Aut)$, where *Goal*, *Init* $\in \mathcal{X}$ are graph class expressions called the *overall goal* and the *initial environment specification*, respectively, and *Aut* is a set of autonomous units.

In a community all units work on the common environment in a self-controlled way by applying their rules. The change relation integrated in the semantics of autonomous units makes it possible to define a parallel semantics of a community in which every autonomous unit may perform its transformation processes. From the point of view of a single autonomous unit, the changes of the environment that are not caused by itself must be activities of the other units in the community. This is reflected in the following definition.

Definition 4 (Change relation). Let $COM = (Goal, Init, Aut)$ be a community. Then for each $aut \in Aut$ the *change relation Change(aut)* w.r.t. aut is given by the parallel rules composed of rules of the autonomous units in COM other than aut as metarules, i.e. $Change(aut) = \bigcup_{aut' \in Aut - \{aut\}} SEM((P_{aut'})_*)$.

Every transformation process of a community must start with a graph specified as an initial environment of the community. Moreover, it must be in the parallel semantics of every autonomous unit participating in the community. Analogously to successful transformation processes of autonomous units, a finite transformation process of a community is successful if its last environment satisfies the overall goal. Every infinite transformation process of a community is successful if it meets infinitely many environments that satisfy the overall goal.

Definition 5 (Parallel community semantics)

1. Let $COM = (Goal, Init, Aut)$. Then the *parallel community semantics* of COM consists of all finite or infinite sequences $s = (G_0, G_1, \ldots) \in SEQ(\mathcal{G})$ such that $G_0 \in SEM(Init)$ and $s \in PAR_{Change(aut)}(aut)$ for all $aut \in Aut$.
2. The sequence s is called a *successful transformation process* if s is finite and $G_{|s|} \in SEM(Goal)$ or there is an infinite monotone sequence $i_0 < i_1 < \cdots$ such that $G_{i_j} \in SEM(Goal)$ for all $j \in \mathbb{N}$.
3. The parallel community semantics is denoted by $PAR(COM)$. Its elements are called parallel processes of COM.

As the definition of the community semantics shows, there is a strong connection between the semantics of a community $COM = (Goal, Init, Aut)$ and the semantics of an autonomous unit $aut \in Aut$. The parallel semantics of COM is a subset of the semantics of aut with respect to the change relation $Change(aut)$. Conversely, one may take the intersection of the parallel semantics of all autonomous units with respect to their own change relation and restrict it to the sequences starting in an initial environment. Then one gets the parallel semantics of the community. This reflects the autonomy because no unit can be forced to do anything that is not admitted by its own control.

Example

In following, we shortly illustrate how communities of autonomous units can be used to find shortest paths by working in parallel. The presented community $CAU(spath)$ is a parallel variant of the famous shortest-path algorithm of Floyd [18].

As initial environments $CAU(spath)$ admits all directed edge-labeled graphs so that every edge from v to v' is labeled with a number representing the distance from v to v'. The set of autonomous units of $CAU(spath)$ consists of the two units *minimum* and *sum* presented in Fig. 1 above.

The goal of the community $CAU(spath)$ is twofold: (1) Whenever there is an edge e and a path p from a node v to a node v' the distance of e has to be less or equal to the distance of p (i.e. the distances edges are the shortest). (2) Whenever there is an edge from a node v to a node v', there is a shortest path from v to v' in the initial environment with the same distance. (This guarantees that the computed edges yield the distances of the shortest paths in the initial graphs.)

The parallel semantics of $CAU(spath)$ is equal to the parallel semantics of the unit *minimum* if the change relation is given by all (parallel) transformations of *sum* and if the transformation processes start with an initial environment of $CAU(spath)$. The analogous property holds for the unit *sum*. More formally, let ID be the identity relation on environments. This relation does not allow environment changes and can be realized with the null rule as the only meta rule. For any set $S \subseteq SEQ(\mathcal{G})$ let $REL(S)$ consist of all pairs (G_0, G_n) for which there is a finite sequence $(G_0, \ldots, G_n) \in S$. Finally, for each autonomous unit aut let $SEM(aut)$ be the semantic relation obtained from all parallel transformations of aut that obey the control condition, i.e. $SEM(aut) = REL(PAR_{ID}(aut))$. Then

$$PAR(CAU(spath)) = PAR_{SEM(minimum)}(sum)|SEM(Init_{CAU(spath)})$$
$$= PAR_{SEM(sum)}(minimum)|SEM(Init_{CAU(spath)}).$$

Moreover it can be shown that the community $CAU(spath)$ works correctly. This means that the parallel semantics of $CAU(spath)$ contains only finite sequences $(G_0, \ldots, G_n)$ such that for every two nodes v and v' there is an edge e with distance x in G_n from v to v' if and only if the shortest path in G_0 from v to v' has distance x (cf. [19] for a correctness proof concerning the sequential variant of this algorithm).

5   Petri Nets

The area of Petri nets (see, e.g., [4,5]) is established as one of the oldest, well-known, and best studied frameworks in which parallelism is precisely introduced and investigated. Hence it is meaningful to relate Petri nets with the parallel semantics of communities of autonomous units and to shed some light on the significance of the latter in this way. It turns out for instance that place/transition nets, which are the most frequently used variants of Petri nets, can be seen as a special case of communities of autonomous units where the transitions play the role of the units.

A place/transition system $S = (P, T, F, m_0)$ consists of a set P of places, a set T of transitions, a flow relation $F \subseteq (P \times T) \cup (T \times P)$, and an initial marking $m_0 : P \to \mathbb{N}$, i.e. $m_0 \in P_*$. The sets P and T are assumed to be disjoint so that $N = (P \cup T, F)$ is a bipartite graph (with the projections as source and target maps respectively).

The firing of enabled transitions transforms markings that are multisets of places. This is formally defined as follows.

A multiset $m \in P_*$ is called a marking. A transition $t \in T$ is enabled w.r.t. m if $^\bullet t \leq m$ where $^\bullet t : P \to \mathbb{N}$ describes the input places of t that flow into t, i.e. $^\bullet t(p) = 1$ if $(p, t) \in F$ and $^\bullet t(p) = 0$ otherwise. The order $^\bullet t \leq m$ is defined place-wise, i.e. $^\bullet t(p) \leq m(p)$ for all $p \in P$ or, in other words, $m(p) \neq 0$ if $(p, t) \in F$. If t is enabled w.r.t. m, it can fire resulting in a marking which is obtained by subtracting $^\bullet t$ from m and by adding $t^\bullet$ given by $t^\bullet(p) = 1$ if $(t, p) \in F$ and $t^\bullet(p) = 0$ otherwise. Such a firing is denoted by $m \, [t\rangle \, m - {}^\bullet t + t^\bullet$. If one interprets $m(p)$ as the number of tokens on the place p, then the firing of t removes one token from each input place of t and puts a new token on each of the output places of t.

Analogously, a multiset of transitions $\tau \in T_*$ can be fired in parallel by summing up all input places and all output places:

$$m \, [\tau\rangle \, m - {}^\bullet\tau + \tau^\bullet \text{ provided that } {}^\bullet\tau \leq m.$$

Here $^\bullet\tau$ and $\tau^\bullet$ are defined by $^\bullet\tau(p) = \sum_{t \in T} \tau(t) * {}^\bullet t(p)$ and $\tau^\bullet(p) = \sum_{t \in T} \tau(t) * t^\bullet(p)$ for all $p \in P$, and the order $^\bullet\tau \leq m$ is again place-wise defined, i.e. $^\bullet\tau(p) \leq m(p)$ for all $p \in P$.

Now one may consider the underlying net, which is the bipartite graph N, together with a marking as an environment. This is represented by the marking because the net is kept invariant. The transitions can be seen as rules and the firing of multisets of transitions as parallel rule application. As environment class expressions, we need single markings describing themselves as initial markings and the constant all accepting all environments. The only control condition needed is the constant $free$ allowing a unit the free choice of rules. Then these components form a graph transformation approach, and a place/transition system $S = (P, T, F, m_0)$ can be translated into a community of autonomous units $CAU(S) = (all, m_0, \{aut(t) \mid t \in T\})$ with $aut(t) = (all, \{t\}, free)$.

A parallel process of $CAU(S)$ is a sequence of markings $m_0 m_1 \ldots$ such that for each two successive markings m_i and m_{i+1}, there is a multiset τ_{i+1} of transitions that is enabled by m_i and yields m_{i+1} if fired. Therefore one gets a firing sequence $m_0 [\tau_1\rangle m_1 [\tau_2\rangle \ldots$. Conversely, given such a firing sequence, one may remove the firing symbols including the multisets of transitions and obtain a parallel process of $CAU(S)$ as parallel rule application coincides with firing of multisets of transitions. This proves that the community of autonomous units $CAU(S)$ mimics the place/transition system S correctly. The following figure depicts the relation.

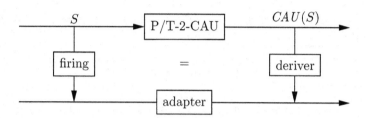

The adapter transforms a firing sequence into a sequence of markings by removing the firing symbol (including the fired multisets of transitions) between each two successive markings.

6   Cellular Automata

Cellular automata (see, e.g., [20]) are well-known computational devices that exhibit massive parallelism. A cellular automaton consists of a network of cells each in a particular state. In a computational step, all cells change their states in parallel depending on the states of their neighbours. To simplify technicalities, one may assume that the neighbourhoods of all cells are regular meaning that they have the same number of neighbours and that the state transition of all cells is based on the same finite-state automaton. This leads to the following formal definition.

A cellular automaton is a system $CA = (G, A, init)$ where

- $G = (V, E, s, t, l)$ is a regular graph of type k subject to the condition: for each $v \in V$, there is a sequence of edges $e(v)_1 \cdots e(v)_k$ with $s(e(v)_i) = v$ and $l(e(v)_i) = i$ for all $i = 1, \ldots, k$,
- $A = (Q, Q^k, d)$ is a finite-state automaton, i.e. Q is a finite set of states, Q^k is the input set and $d \subseteq Q \times Q^k \times Q$ is the state transition with k-tuples of states as inputs, and
- $init: V \to Q$ is the initial configuration.

If the graph G is infinite, one assumes a sleeping state $q_0 \in Q$ in addition such that $d(q_0, q_0^k) = \{q_0\}$ and $active(init) = \{v \in V \mid init(v) \neq q_0\}$ is finite.

The latter means that only a finite number of nodes is not sleeping initially and that the sleeping state can only wake up if not all inputs are sleeping.

The edge sequence $e(v)_1 \cdots e(v)_k$ yields the neighbours of v as targets, i.e. $t(e(v_1)) \cdots t(e(v)_k))$.

A configuration is a mapping $con: V \to Q$ that assigns each node (which represent cells) an actual state. Configurations can be updated by state transitions of all actual states using the states of the neighbours as input.

Let $con: V \to Q$ be a configuration. Then $con': V \to Q$ is a directly derived configuration, denoted by $con \vdash con'$, if the following holds for every $v \in V$:

$$con'(v) \in d(con(v), con(t(e(v)_1)) \cdots con(t(e(v)_k))).$$

The semantics of a cellular automaton CA is given by all configurations that can be derived from the initial configuration:

$$L(CA) = \{con \mid init \vdash^* con\}$$

It is worth noting and easy to prove that all configurations derivable from the initial configuration have a finite number of nodes with non-sleeping states. Typical examples of regular graphs underlying cellular automata are the following: The set of nodes is the set of all points in the plane with integer coordinates, i.e. $\mathbb{Z} \times \mathbb{Z}$. Then there are various choices for the neighbourhood of a node $(x, y) \in \mathbb{Z} \times \mathbb{Z}$. that establish the set of edges with sources and targets. Typical ones are:

1. the four nearest nodes (to the north, east, south and west): $(x, y + 1), (x + 1, y), (x, y - 1), (x - 1, y)$,
2. the eight nearest nodes: $(x, y+1), (, x+1, y+1), (x+1, y), (x+1, y-1), (x, y-1), (x - 1, y - 1), (x - 1, y), (x - 1, x + 1)$,
3. only the neighbours to the south and the west: $(x, y - 1), (x - 1, y)$.

The edges connecting a node with a neighbour may be numbered in the given order.

Cellular automata can be translated into communities of autonomous units where each cell is transformed into a unit.

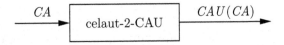

The environments are given by the configurations. To get a graph representation of a configuration con, the underlying regular graph G is extended by a loop at each node v which is labeled with $con(v)$, i.e. $(G, con) = (V, E + V, \bar{s}, \bar{t}, \bar{l})$, such that G is a subgraph and $\bar{s}(v) = \bar{t}(v) = v$ and $\bar{l}(v) = con(v)$ for all $v \in V \subseteq E + V$.

The community of autonomous units $CAU(CA)$ associated with a cellular automaton $CA = (G, A, init)$ gets $(G, init)$ as initial environment and an autonomous unit $aut(v)$ for each $v \in V$.

Each of these units has the same rules with positive context which reflect the state transition:

provided that $q' \in d(q, q_1, \cdots, q_k)$ and not all the states $q, q_1, \ldots, q_k$ are sleeping. Moreover each unit $aut(v)$ has got a control condition requiring that the central node must be mapped to v. This means that the matching of the left-hand side of each rule is fixed and no search for it is needed. Moreover, the matchings of rules of different units are not overlapping so that the rules can be applied in parallel. If a node is sleeping and all its neighbours are sleeping too, then no rule can be applied. A parallel rule is maximal if all other nodes are matched. According to this construction, the application of such a maximal parallel rule to the environment (G, con) yields an environment (G, con') such that $con \longmapsto con'$. This means that the application of a maximal parallel rule corresponds exactly to a derivation step on the respective configurations.

In other words, the semantics of a cellular automaton CA and the parallel semantics $PAR(CAU(CA))$ of the community of autonomous units $CAU(CA)$ are nicely related to each other if one applies maximal parallel rules only. Let $L(PAR(CAU(CA)))$ be the set of configurations con such that a parallel process $(G, init) \cdots (G, con) \in PAR(CAU(CA))$ exists. Then $L(PAR(CAU(CA)))$ equals $L(CA)$. This correctness result is depicted by the following figure.

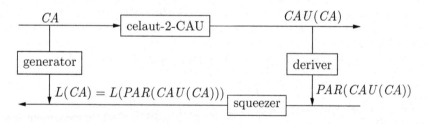

A finite-state automaton fitting the third neighbourhood is

$$SIER = (\{b, w\}, \{b, w\}^2, d)$$

with $d(b, x, y) = b$ for all $x, y \in \{b, w\}, d(w, b, w) = d(w, w, b) = b$, and $d(w, b, b) = d(w, w, w) = w$. The state w is sleeping.

The initial configuration may map the node $(0, 0)$ to b and all others to w.

There is a very nice pictorial interpretation of this cellular automaton. Each node (x, y) is represented by the square spanned by the points $(x, y), (x, y + 1), (x + 1, y + 1), (x + 1, y)$. If a configuration con assigns b to (x, y), the square gets the color black and white otherwise. The initial configuration consists of a single black square. Because the automaton is deterministic, there is exactly one derivation for each length, where the shorter derivations are initial sections of the longer ones. The first five steps are

After 15 steps the picture looks as follows:

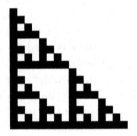

And all derived configurations can be seen as approximations of the Sierpinski triangle, a famous fractal. (see, e.g., [21]).

7   Multiagent Systems

Multiagent systems are modelling and programming devices well-known in artificial intelligence (see, e.g., Wooldridge et al. [6]). A multiagent system provides a set of agents and an initial environment state. Starting at this state, the agents change environment states step by step where they act together in parallel in each step. Each agent can perceive the current environment state at least partly. Based on this perception and its own intention, the agent chooses an action to be performed next. Therefore, a process in a multiagent system MAS is a sequence

$$es_0 \ es_1 \ es_2 \ \cdots$$

of environment states es_i for all i where es_0 is initial. Each environment state es_{i+1} is given by the state transition τ of MAS depending on the previous state es_i and the action $act(ag)_i$ chosen by every agent ag of MAS. The choice of such an action is made according to the function do_{ag} each agent ag is provided with. The do-function yields an action depending on the agent's perception $perceive_{ag}(es_i)$ of the current state and the agent's intention $intend_{ag}$. The global state transition τ and the functions $do_{ag}, perceive_{ag}$ and $intend_{ag}$ which are individually assigned to each agent ag of MAS are assumed to satisfy some consistency properties (cf. [6] for details). Altogether, multiagent systems form a logical and axiomatic approach to model distributed information processes that interact on common environment states. It should be noted that all functions of

MAS are allowed to be nondeterministic so that chosen actions as well as the next state may not be uniquely determined.

Communities of autonomous units are nicely related to multiagent systems as may be not too surprising from the description above. Actually, a community of autonomous units $CAU = (Goal, Init, Aut)$ turns out to be a particular rule-based model of multiagent systems. The environment states are the environment graphs. The agents are the units. The initial graphs are explicitly given. The rules – or the parallel rules likewise – of a unit are the actions of the agent embodied by the unit. The control condition plays the role of the do-function because it identifies the rules that are allowed to be applied next. As the control condition can take into account the current environment graph, the perception of the agent is also reflected. The most important aspect of the correspondence between agents and units is the transition function that is made operational by means of parallel rule application. The parallel rule to be applied in each step is just the sum of all rules chosen by the various units according to their control. If one considers the parallel rules of a unit as actions, the parallel processes of the community and the processes of the corresponding multiagent system coincide. If only the rules are actions, the multiagent system is not parallel with respect to single agents. That all agents must act in parallel in each step is a minor difference to community processes because a multiagent system may provide void actions without effect to the environment.

The relation between communities of autonomous units and multiagent systems is only sketched because a full formal treatment is beyond the scope of the paper. But even on this informal level, it should be clear that both concepts fit nicely together and may profit from each other. Communities of autonomous units represent explicit models of multiagent systems on one abstract, implementation-independent level with a precise, rule-based operational semantics. The *perceive-do* mechanism of multiagent systems to choose next actions provides a wealthy supply of control conditions that can be employed in modeling by means of autonomous units.

8   Conclusion

In this paper, we have supplemented the sequential-process semantics of autonomous units in [1] by a parallel-process semantics which allows the units of a community to act and interact simultaneously in a common environment. Moreover, we have studied the relationship of autonomous units to three other modeling frameworks that provide notions of parallelism: Petri nets, cellular automata, and multiagent systems. While the first two have been correctly transformed into autonomous units, autonomous units have turned out to be models of multiagent systems in that the environments are instantiated as graphs, the actions of agents as rules, and the environment transformation as parallel rule application. This is the very first step of the investigation of autonomous units in a parallel setting. The future study may include the following topics:

1. The interplay of different autonomous units in a community should be further investigated on a theoretical as well as on a case-study based level. In particular, it would be interesting to check under which conditions the result concerning the semantics of the units *sum* and *minimum* in the example of Section 4 can be generalized to arbitrary communities with more than two units.
2. Besides Petri nets, the theory of concurrency offers a wide spectrum of notions of processes like communicating sequential processes, calculus of communicating systems, traces, and bigraphs. A detailed comparison of them with autonomous units can lead to interesting insights.
3. The basic idea of autonomous units is that each of them decides for itself which rule is to be applied next. They are independent of each other and the parts of the environment graphs where their rules apply may be far away from each other. Hence a sequential behaviour of the community (like in many card and board games) will be rarely adequate. But also the parallel behaviour does not always reflect the actual situations to be modeled because a parallel step provides a graph before and a graph after the step whereas there may be activities of units that cannot be related to each other with respect to time. A proper concurrent semantics of autonomous units may fix this problem.
4. In all explicit examples, we have made use of the fact that independent rule applications can be applied in parallel. This holds in the DPO approach (as well as in the SPO approach) together with several other properties and constructions that relate parallel and sequential processes yielding true concurrency for example. It seems to be meaningful to extend these considerations to the framework of autonomous units.

References

1. Hölscher, K., Kreowski, H., Kuske, S.: Autonomous units and their semantics — the sequential case. In: Proc. International Conference of Graph Transformation. Volume 4178 of Lecture Notes in Computer Science. (2006) 245–259
2. Kuhn, A.: Prozessketten – Ein Modell für die Logistk. In Wiesendahl, H.P., ed.: Erfolgsfaktor Logistikqualität. Springer Verlag (2002) 58–72
3. Scheer., A.W.: Vom Geschäftsprozeß zum Anwendungssystem. Springer Verlag (2002)
4. Reisig, W.: Elements of Distributed Algorithms: Modeling and Analysis With Petri Nets. Springer Verlag (1998)
5. Bause, F., Kritzinger, P.: Stochastic Petri Nets - An Introduction to the Theory (Second Edition). Vieweg & Sohn Verlag Braunschweig/Wiesbaden (Germany) (2002)
6. Wooldridge, M., Jennings, N.R.: Intelligent agents: Theory and practice. The Knowledge Engineering Review **10**(2) (1995)
7. Kennedy, J., Eberhart, R.C.: Swarm Intelligence. Morgan Kaufmann (2001)
8. Rozenberg, G., Salomaa, A., eds.: Handbook of Formal Languages, Vol. 1–3. Springer Verlag (1997)

9. Baader, F., Nipkow, T.: Term Rewriting and All That. Cambridge University Press (1998)
10. Rozenberg, G., ed.: Handbook of Graph Grammars and Computing by Graph Transformation, Vol. 1: Foundations. World Scientific, Singapore (1997)
11. Ehrig, H., Engels, G., Kreowski, H.J., Rozenberg, G., eds.: Handbook of Graph Grammars and Computing by Graph Transformation, Vol. 2: Applications, Languages and Tools. World Scientific, Singapore (1999)
12. Ehrig, H., Kreowski, H.J., Montanari, U., Rozenberg, G., eds.: Handbook of Graph Grammars and Computing by Graph Transformation, Vol. 3: Concurrency, Parallelism, and Distribution. World Scientific, Singapore (1999)
13. Păun, G., Rozenberg, G., Salomaa, A.: DNA Computing — New Computing Paradigms. Springer Verlag (1998)
14. Kreowski, H.J., Kuske, S.: Graph transformation units with interleaving semantics. Formal Aspects of Computing **11**(6) (1999) 690–723
15. Andries, M., Engels, G., Habel, A., Hoffmann, B., Kreowski, H.J., Kuske, S., Plump, D., Schürr, A., Taentzer, G.: Graph transformation for specification and programming. Science of Computer Programming **34**(1) (1999) 1–54
16. Habel, A., Heckel, R., Taentzer, G.: Graph grammars with negative application conditions. Fundamenta Informaticae **26**(3,4) (1996) 287–313
17. Corradini, A., Ehrig, H., Heckel, R., Löwe, M., Montanari, U., Rossi, F.: Algebraic approaches to graph transformation part I: Basic concepts and double pushout approach. [10] 163–245
18. Even, S.: Graph Algorithms. Computer Science Press (1979)
19. Kreowski, H.J., Kuske, S.: Graph transformation units and modules. In Ehrig, H., Engels, G., Kreowski, H.J., Rozenberg, G., eds.: Handbook of Graph Grammars and Computing by Graph Transformation, Vol. 2: Applications, Languages and Tools. World Scientific, Singapore (1999) 607–638
20. Wolfram, S.: A New Kind of Science. Wolfram Media, Inc. (2002)
21. Peitgen, H., Jürgens, H., Saupe, D.: Chaos and Fractals. Springer, Berlin (2004)

Reasoning Support for CASL
with Automated Theorem Proving Systems*

Klaus Lüttich[1] and Till Mossakowski[2]

[1] SFB/TR8, University of Bremen
[2] University of Bremen and DFKI Lab Bremen
{luettich,till}@informatik.uni-bremen.de

Abstract. We connect the algebraic specification language CASL with
a variety of automated first-order provers. The heart of this connection
is an institution comorphism from CASL to *SoftFOL* (softly typed first-
order logic); the latter is then translated to the provers' input syntaxes.
We also describe a GUI integrating the translations and the provers
into the Heterogeneous Tool Set. We report on experiences with provers,
which led to fine-tuning of the translations. This framework can also be
used for checking consistency of specifications.

1 Introduction

The Common Algebraic Specification Language (CASL) [3,5] is a modern standard for axiomatic specification using first-order logic and datatypes. During
the development of CASL specifications, it is crucial to have good proof support – be it for proving intended consequences of specifications, or checking their
consistency.

So far, only interactive provers like Isabelle [12] have been connected to CASL.
This paper describes reasoning support for CASL with automated first-order logic
theorem proving (ATP) systems made available through the Heterogeneous Tool
Set (HETS) [11]. Although ATP systems do not provide reasoning support for
all features of CASL, they can take us surprisingly far.

The ATP reasoning support for CASL described in this paper is based on softly
typed first-order logic (*SoftFOL*) which is presented as an institution. Furthermore, a coding of a sublogic of CASL into *SoftFOL* is formalized as an institution
comorphism. Two different syntactical representations (DFG and TPTP) of *Soft-
FOL* allow the connection of SPASS [15] and MathServe [16] with HETS. While
SPASS is an ATP system by itself, MathServe offers reasoning support for different ATP systems as web services and a broker service, which chooses a suitable
ATP system after classification of the problem. Furthermore, the integration
of reasoning support into HETS is presented together with a description of the
graphical user interface (GUI) used to control the ATP. Finally, we summarize
optimisations of the coding implemented in HETS.

---

* This paper has been supported by Deutsche Forschungsgemeinschaft in the
SFB/TR8 "Spatial Cognition" and in the project MULTIPLE under grant KR
1191/5-2.

J.L. Fiadeiro and P.-Y. Schobbens (Eds.): WADT 2006, LNCS 4409, pp. 74–91, 2007.
© Springer-Verlag Berlin Heidelberg 2007

2   Institutions and Their Comorphisms

The use of automated first-order provers for CASL requires translations to be set up between the CASL logic and the provers' logics. We formalize these translations as *institution comorphisms*. Institutions were introduced by Goguen and Burstall [7] to capture the notion of logical system and abstract away from the details of signatures, sentences, models and satisfaction.

Let $\mathcal{CAT}$ be the category of categories and functors.[1]

Definition 1. *An institution* $I = (\mathbf{Sign}, \mathbf{Sen}, \mathbf{Mod}, \models)$ *consists of*

- *a category* **Sign** *of signatures,*
- *a functor* $\mathbf{Sen}\colon \mathbf{Sign} \longrightarrow \mathbf{Set}$ *giving, for each signature* Σ, *the set of sentences* $\mathbf{Sen}(\Sigma)$, *and for each signature morphism* $\sigma\colon \Sigma \longrightarrow \Sigma'$, *the sentence translation map* $\mathbf{Sen}(\sigma)\colon \mathbf{Sen}(\Sigma) \longrightarrow \mathbf{Sen}(\Sigma')$, *where often* $\mathbf{Sen}(\sigma)(\varphi)$ *is written as* $\sigma(\varphi)$,
- *a functor* $\mathbf{Mod}\colon \mathbf{Sign}^{op} \longrightarrow \mathcal{CAT}$ *giving, for each signature* Σ, *the category of models* $\mathbf{Mod}(\Sigma)$, *and for each signature morphism* $\sigma\colon \Sigma \longrightarrow \Sigma'$, *the reduct functor* $\mathbf{Mod}(\sigma)\colon \mathbf{Mod}(\Sigma') \longrightarrow \mathbf{Mod}(\Sigma)$, *where often* $\mathbf{Mod}(\sigma)(M')$ *is written as* $M'|_{\sigma}$,
- *a satisfaction relation* $\models_{\Sigma} \subseteq |\mathbf{Mod}(\Sigma)| \times \mathbf{Sen}(\Sigma)$ *for each* $\Sigma \in |\mathbf{Sign}|$,

such that for each $\sigma\colon \Sigma \longrightarrow \Sigma'$ *in* **Sign** *the following* satisfaction condition *holds:*

$$M' \models_{\Sigma'} \sigma(\varphi) \Leftrightarrow M'|_{\sigma} \models_{\Sigma} \varphi$$

for each $M' \in |\mathbf{Mod}(\Sigma')|$ *and* $\varphi \in \mathbf{Sen}(\Sigma)$, *expressing that truth is invariant under change of notation and enlargement of context.*   □

A *theory* in an institution is a pair $T = (\Sigma, \Gamma)$ consisting of a signature $Sign(T) = \Sigma$ and a set of Σ-sentences $Ax(T) = \Gamma$, the axioms of the theory. *Theory morphisms* are signature morphisms that map axioms to logical consequences.

Institution comorphisms [6] allow the expression of the fact that one institution I is *included* or *encoded* into an institution J. Given institutions I and J, an *institution comorphism* $\rho = (\Phi, \alpha, \beta)\colon I \longrightarrow J$ consists of

- a functor $\Phi\colon \mathbf{Sign}^{I} \longrightarrow \mathbf{Sign}^{J}$,
- a natural transformation $\alpha\colon \mathbf{Sen}^{I} \longrightarrow \mathbf{Sen}^{J} \circ \Phi$,
- a natural transformation $\beta\colon \mathbf{Mod}^{J} \circ \Phi^{op} \longrightarrow \mathbf{Mod}^{I}$

such that the following *satisfaction condition* is satisfied for all $\Sigma \in \mathbf{Sign}^{I}$, $M' \in \mathbf{Mod}^{J}(\Phi(\Sigma))$ and $\varphi \in \mathbf{Sen}^{I}(\Sigma)$:

$$M' \models^{J}_{\Phi(\Sigma)} \alpha_{\Sigma}(\varphi) \Leftrightarrow \beta_{\Sigma}(M') \models^{I}_{\Sigma} \varphi.$$

In more detail, this means that each signature $\Sigma \in \mathbf{Sign}^{I}$ is translated to a signature $\Phi(\Sigma) \in \mathbf{Sign}^{J}$, and each signature morphism $\sigma\colon \Sigma \longrightarrow \Sigma' \in \mathbf{Sign}^{I}$ is

---

[1] Strictly speaking, $\mathcal{CAT}$ is not a category but only a so-called quasicategory, which is a category that lives in a higher set-theoretic universe.

translated to a signature morphism $\Phi(\sigma)\colon \Phi(\Sigma)\longrightarrow \Phi(\Sigma')\in \mathbf{Sign}^J$. Moreover, for each signature $\Sigma \in \mathbf{Sign}^I$, we have a sentence translation map $\alpha_\Sigma\colon \mathbf{Sen}^I(\Sigma)\longrightarrow \mathbf{Sen}^J(\Phi(\Sigma))$ and a model translation functor $\beta_\Sigma : \mathbf{Mod}^J(\Phi(\Sigma))\longrightarrow \mathbf{Mod}^I(\Sigma)$.

A *simple theoroidal comorphism* is like a comorphism, except that the signature translation functor Φ ends in the category of theories over the target institution.

3 The CASL Logic

CASL, the Common Algebraic Specification Language, has been designed by CoFI, the international *Common Framework Initiative for algebraic specification and development* [1], with the goal to subsume many previous algebraic specification languages and to provide a standard language for the specification and development of modular software systems. See the CASL user manual [3] and reference manual [5] for further information.

Here, we concentrate on CASL *basic specifications*, designed for writing single specification modules. CASL also provides constructs for structured and architectural specifications and specification libraries.

The logic of CASL basic specifications combines first-order logic and induction (the latter is expressed using so-called sort generation constraints, and is needed for the specification of the usual inductive datatypes) with subsorts and partial functions. The institution underlying CASL is introduced in two steps [5]: first, we introduce many-sorted partial first-order logic with sort generation constraints and equality ($PCFOL^=$), and then, subsorted partial first-order logic with sort generation constraints and equality ($SubPCFOL^=$) is described in terms of $PCFOL^=$.

3.1 Partial First-Order Logic

We now sketch the institution $PCFOL^=$ of many-sorted partial first-order logic with sort generation constraints and equality. Full details can be found in [10,5].

A *many-sorted* CASL *signature* $\Sigma = (S, TF, PF, P)$ consists of a set S of sorts, two $S^* \times S$-indexed[2] sets $TF = (TF_{w,s})$ and $PF = (PF_{w,s})$ of total and partial operation symbols, and an S^*-indexed set $P = (P_w)$ of predicate symbols. Function and predicate symbols are written $f : \bar{s} \to t$ and $p : \bar{s}$, respectively, where t is a sort and $\bar{s}$ is a list $s_1 \ldots s_n$ of sorts, thus determining their *name* and *profile*. Symbols with identical names are said to be *overloaded*; they may be referred to by just their names in CASL specifications, but are always qualified by profiles in fully statically analysed sentences. Signature morphisms map the sorts and the function and predicate symbols in a compatible way, such that the totality of function symbols is preserved.

Models are many-sorted partial first order structures, interpreting sorts as carrier sets, total (partial) function symbols as total (partial) functions and predicate symbols as relations. Homomorphisms between such models are so-called

---

[2] S^* is the set of strings over S.

weak homomorphisms. That is, they are total as functions, and they preserve (but not necessarily reflect) the definedness of partial functions and the satisfaction of predicates.

Concerning *reducts*, if $\sigma: \Sigma_1 \longrightarrow \Sigma_2$ is a signature morphism and M is a Σ_2-model, then $M|_\sigma$ is the Σ_1-model which interprets a symbol by first translating it along σ and then taking M's interpretation of the translated symbol. Reducts of homomorphisms are defined similarly.

Given a many-sorted signature $\Sigma = (S, TF, PF, P)$ and a pairwise disjoint S-indexed set of variables X, the set $T_\Sigma(X)$ of *terms* over Σ and X is defined inductively as usual, using variables and operation symbols. Concerning the semantic interpretation of terms in a model, variable assignments are total, but the value of a term w.r.t. a variable assignment may be undefined, due to the application of a partial function during the evaluation of the term. Undefinedness propagates from subterms to superterms.

Sentences are built from atomic sentences using the usual features of first order logic. Given Σ and X, the set $AF_\Sigma(X)$ of *many-sorted atomic Σ-formulas* with variables in X contains:

1. $p_w(t_1, \ldots, t_n)$, for $t_i \in T_\Sigma(X)_{s_i}, p \in P_w, w = s_1 \ldots s_n \in S^*$,
2. $t \overset{e}{=} t'$, for $t, t' \in T_\Sigma(X)_s, s \in S$ (existential equations),
3. $t = t'$, for $t, t' \in T_\Sigma(X)_s, s \in S$ (strong equations),
4. *def* t, for $t \in T_\Sigma(X)_s, s \in S$ (definedness assertions).

A definedness assertion holds w.r.t. a given valuation in a model if the term is defined under that valuation. A strong equation holds if its two sides are both defined or both undefined under the valuation, and in case of definedness, they are interpreted equally. An existence equation holds if both sides are defined and interpreted equally. A predicate application holds if all the terms are defined under the given valuation, and the resulting tuple of model elements is in the corresponding predicate. In this way, we retain the simplicity of a two-valued logic.

The satisfaction of compound formulas, built from atomic formulas using logical connectives and quantifiers, is defined as usual in first-order logic.

There is an additional type of sentence that goes beyond first-order logic: a *sort generation constraint* states that a given set of sorts is generated by a given set of functions, i.e. that all the values of the generated sorts are reachable by some term in the function symbols, possibly containing variables of other sorts.

Formally, a sort generation constraint over a signature Σ is a triple (S', F', θ), where $\theta: \bar{\Sigma} \longrightarrow \Sigma$, $\bar{\Sigma} = (\bar{S}, \bar{TF}, \bar{PF}, \bar{P})$, $S' \subseteq \bar{S}$ and $F' \subseteq \bar{TF} \cup \bar{PF}$. A Σ-constraint (S', F', θ) is satisfied in a Σ-model M if the carriers of $M|_\theta$ of the sorts in S' are generated by the function symbols in F', i.e. for every sort $s \in S'$ and every value $a \in (M|_\theta)_s$, there is a $\bar{\Sigma}$-term t containing only function symbols from F' and variables of sorts not in S' such that $\nu^\#(t) = a$ for some assignment ν into $M|_\theta$.

Translation of a sentence along a signature morphism just replaces all the symbols in the sentence according to the signature morphism; this (together with the reducts) fulfills the satisfaction condition [10]. Sort generation constraints cannot be translated in this way; instead, the extra signature morphism component is used: The translation of a constraint (S', F', θ) along σ is $(S', F', \sigma \circ \theta)$. This obviously leads to fulfillment of the satisfaction condition.

This completes the definition of the institution $PCFOL^=$.

CASL additionally has unique existential quantification and a conditional term construct. These are coded out; see Sect. 7 for details.

3.2   Subsorted Partial First-Order Logic

Subsorted partial first-order logic is defined in terms of partial first-order logic. The basic idea is to reduce subsorting to injections between sorts. While in the subsorted institution, these injections have to occur explicitly in the sentences, in the CASL language, they may be left implicit. Apart from the injections, one also has partial projection functions (one-sided inverses of the injections) and membership predicates.

The institution $SubPCFOL^=$ is defined as follows, extending the notion of order-sorted signatures as given by Goguen and Meseguer [8].

A *subsorted signature* $\Sigma = (S, TF, PF, P, \leq_S)$ consists of a many-sorted signature (S, TF, PF, P) together with a reflexive transitive *subsort relation* $\leq_S$ on the set S of sorts.

For a subsorted signature, $\Sigma = (S, TF, PF, P, \leq_S)$, we define *overloading relations* (also called *monotonicity orderings*), $\sim_F$ and $\sim_P$, for function and predicate symbols, respectively:

Let $f : w_1 \longrightarrow s_1, f : w_2 \longrightarrow s_2 \in TF \cup PF$, then

$$f : w_1 \longrightarrow s_1 \sim_F f : w_2 \longrightarrow s_2$$

iff there exist $w \in S^*$ with $w \leq w_1$ and $w \leq w_2$ and $s \in S$ with $s_1 \leq s$ and $s_2 \leq s$. Let $p : w_1, p : w_2 \in P$, then $p : w_1 \sim_P p : w_2$ iff there exists $w \in S^*$ with $w \leq w_1$ and $w \leq w_2$.

A *signature morphism* $\sigma : \Sigma \rightarrow \Sigma'$ is a many-sorted signature morphism that preserves the subsort relation and the overloading relations.

With each subsorted signature $\Sigma = (S, TF, PF, P, \leq_S)$ we associate a many-sorted signature $\hat{\Sigma}$, which is the extension of the underlying many-sorted signature (S, TF, PF, P) with new symbols,

- a total *injection* function symbol $\mathtt{inj} : s \rightarrow s'$, for each pair of sorts $s \leq_S s'$,
- a partial *projection* function symbol $\mathtt{pr} : s' \rightarrow? s$, for each pair of sorts $s \leq_S s'$, and
- a unary *membership* predicate symbol $\in^s : s'$, for each pair of sorts $s \leq_S s'$.

Subsorted Σ-*models* are ordinary many-sorted $\hat{\Sigma}$-models satisfying the following properties (which can be formalized as a set of conditional axioms):

- Embedding operations are total and injective; projection operations are partial, and injective when defined.

- The embedding of a sort into itself is the identity function.
- All compositions of embedding operations between the same two sorts are equal functions.
- Embedding followed by projection is the identity function; projection followed by embedding is included in the identity function.
- Membership in a subsort holds just when the projection to the subsort is defined.
- Embedding is compatible with those operations and predicates that are in the overloading relations.

Signature morphisms, homomorphisms, reducts, sentences, sentence translation and satisfaction are simply inherited via the translation $\Sigma \mapsto \hat{\Sigma}$ from $PCFOL^=$.

This completes the definition of the institution $SubPCFOL^=$.

Every CASL basic specification SP generates, along with its $SubPCFOL^=$-signature Σ, a set Γ of Σ-sentences; together, these determine the *theory* (Σ, Γ) generated by SP. Note that Γ contains not only explicitly stated sentences, but also sentences that are generated e.g. by CASL's powerful datatype constructs (see below), such as the statement that selectors are one-sided inverses of their constructor. See [5] for further details.

3.3 Derived CASL Sublogics

Let $SubCFOL^=$ be the restriction of $SubPCFOL^=$ to signatures without partial functions symbols, and $SulPCFOL^=$ be the restriction of $SubPCFOL^=$ to signatures with a locally filtered subsort relation. Furthermore, $SulCFOL^=$ is the similar restriction of $SubCFOL^=$. Recall that a pre-order is locally filtered, if each connected pair has an upper bound.

4   SPASS, MathServe and *SoftFOL*

This section introduces the ATP system SPASS and the MathServe-system. The latter provides web services for different FOL ATP systems, as well as a broker for FOL ATP web services. The following part of this section introduces the logic *SoftFOL* which provides softly typed FOL, while the last two parts introduce the input languages of SPASS and MathServe.

4.1 SPASS

SPASS [15] is a saturation based automated theorem prover and supports full sorted first-order logic with equality. It has been developed as an open source tool (GPL) since 1991 at the Max Planck Institut Informatik in Saarbrücken, Germany by Christoph Weidenbach, Thomas Hillenbrand, Dalibor Topić et al.

The reasoning methods utilized by SPASS include the following:

- superposition calculus,
- specific inference/reduction rules for sorts,

- splitting rules for explicit case analysis,
- sophisticated clause normal form (CNF) translation.

4.2 MathServe

The MathServe system has been developed by Jürgen Zimmer as open source software (GPL) and provides web services for ATP [16]. MathServe uses state-of-the-art Semantic Web technologies for describing the input, output and effects of the provided reasoning web services. It uses standardised protocols and formats to communicate with client software systems such as HTTP, SOAP and XML.

Table 1. ATP systems provided as web services by MathServe

ATP System	Version	Suitable Problem Classes[a]
DCTP	10.21p	effectively propositional
EP	0.91	effectively propositional; real first-order, no equality; real first-order, equality;
Otter	3.3	real first-order, no equality;
SPASS	2.2	effectively propositional; real first-order, no equality; real first-order, equality
Vampire	8.0	effectively propositional; pure equality, equality clauses contain non-unit equality clauses; real first-order, no equality, non-Horn;
Waldmeister	704	pure equality, equality clauses are unit equality clauses

[a] The list of problem classes for each ATP system is not exhaustive, but only the most appropriate problem classes are named according to benchmark tests made with MathServe by Jürgen Zimmer.

The reasoners provided by MathServe (version 0.81) as web services are summarized in Table 1 (see [14] for further details – all listed ATP systems participated in the competition CASC-20). Additionally, a broker service is provided by MathServe which classifies a given reasoning problem given in FOL with equality and calls the most appropriate reasoning service.

4.3 *SoftFOL*

We now capture the logic underlying the SPASS theorem prover. The institution *SoftFOL* [3] (softly typed first order logic) is a softly (i.e. semantically) typed variant of unsorted (i.e. single-sorted) first-order logic. In the *signatures*, it provides sorts, predicates and total functions. Each predicate and function symbol may optionally have a type profile in terms of the sorts. Overloading of predicates

---

[3] *SoftFOL* it similar to membership equational logic [9], but provides only one (implicit) kind and has operation symbols as part of the signatures, instead of coding them as axioms.

and functions is only allowed if the kind of symbol (predicate or function) and the arities are the same. Subsorting is available as in CASL except that only a locally filtered subsort relation is allowed. *Signature morphisms* are similar to those of CASL; they have to preserve the typing, if present.

SoftFOL models have only one carrier set. Sorts are interpreted as subsets of the carrier. Subsorts must lead to subset inclusions. Each operation or predicate symbol is interpreted as one operation or predicate over the whole carrier. Each typing of an operation leads to the restriction that the operation takes arguments from the subsets as determined by the typing to results as determined by typing.

Sentences in *SoftFOL* are closed untyped FOL sentences, and hence applications of predicates and functions are not qualified with types and a type-correct usage of predicates and functions is not statically checked. (Only an invoked ATP may find incorrect applications to be inconsistent.) Variables may be typed (as operations may be), but again the typing information is not used for sentence formation. Sentences may also involve sort membership tests (as in CASL). *Satisfaction* is mostly as in untyped first-order logic – only the typing of variables leads to a restriction of their possible valuations. Sort generation constraints are available as sentences as in CASL, and their satisfaction is also inherited from CASL (note that this, unlike the case in the rest of *SoftFOL*, involves correctly strongly typed terms only!).

4.4 DFG Syntax

The input language of SPASS is called DFG and is a notation for *SoftFOL*. The DFG format distinguishes three relevant sections of a problem: (1) a list of symbols; (2) a list of declarations; and (3) two lists of formulas for axioms and conjectures. In the list of symbols the arities for functions and predicates are declared and the symbols for sorts are fixed. The list of declarations is a special form of axioms dealing with information about subsorting and (free) generatedness of sorts and the types of predicates and functions. Internally subsort and function type declarations are treated as axioms by SPASS and the other declarations are used for the Knuth-Bendix ordering of symbols. The lists of formulas allow for the typing of variables with sort predicates at the quantification level, but internally the typing of variables is treated as the antecedent of an implication where the quantified formula is the consequent. A symbol declared as sort, predicate or function cannot be used as a variable symbol. The DFG language has a built-in special predicate for equality.

4.5 TPTP Syntax

The TPTP (Thousands of Problems for Theorem Provers) language was invented as a uniform exchange language for ATP systems and is used for the TPTP library of logical problems [13]. This library forms the basis for the annual CADE ATP System Competition (CASC) [14] at the Conference on Automated Deduction (CADE). It is also used as the input language for the MathServe web-services (Sect. 4.2). The TPTP format provides only untyped FOL with an equality predicate. So, there are no constructs for the declaration or typing of symbols. A TPTP problem consists simply of a list of labeled axioms and conjectures.

5   Coding of Logics

The process of coding CASL into the input languages of SPASS and MathServe is performed in three steps, where the first step may be omitted if the CASL theory has no partial functions:

1. Coding of $SulPCFOL^=$ into $SulCFOL^=$, using a comorphism,
2. Coding of $SulCFOL^=$ into $SoftFOL$, using a comorphism,
3. Coding of $SoftFOL$ into DFG or TPTP, using a syntax translation.

The first translation has been described as translation $(5a')$ in [10], modulo the – here inessential – Sub versus Sul. Note that this translation totalizes not only the user-declared partial function, but also uses of partial projection symbols (for the latter, total projection symbols are introduced). The second translation will be detailed below. The translations in the third item are not described as comorphisms, because the logic remains essentially the same and only the syntax changes, while usually a theory (with proof goals) is translated as a whole.

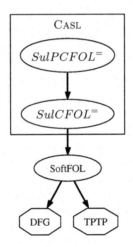

Fig. 1. Relation of the logics and translations

Figure 1 shows the codings as arrows and the logics and syntax formats used for automated theorem proving. The rest of this section covers the second and third step of the list shown above.

5.1   Coding CASL into *SoftFOL*

We now describe a simple theoroidal institution comorphism from $SulCFOL^=$ to *SoftFOL*.

Signatures. CASL signatures are mapped to *SoftFOL* signatures where each CASL symbol is qualified with its kind (sort, pred, op) and its arity, e.g. sort s becomes sort_s and operation $f : s * s \rightarrow r$ becomes op_f_2. The subsorting relation is kept, but for each subsorting relation $s < t$, an operation op_inj: $s \longrightarrow t$ is added (for potential use in sort generation constraints), axiomatized to be the identity

$$\forall x : s . \; \text{op\_inj}(x) = x$$

For each sort in the CASL signature an axiom is generated stating that the sort is not empty.

Models. *SoftFOL* models have only one carrier set. In order to obtain a CASL model, subsets of this carrier set (according to the interpretation of sorts in *SoftFOL*) are taken as carrier sets of a CASL model. Since CASL operations are translated to typed operations in *SoftFOL*, their restrictions are well-defined total operations on the subsets corresponding to the carriers of the CASL model. Subsort injections are interpreted as identities and membership as subset membership. This automatically ensures that the subsorting axioms (see Sect. 3.2) hold.

Sentences. Sentences are translated by erasing all subsorting injection operations; only in sort generation constraints are injections kept (therefore, we needed to introduce special injections into the signatures above).

Satisfaction. The satisfaction condition is clear from the fact that erasing all subsorting injection operations in the sentences corresponds to interpreting them as identities. Note that while *SoftFOL*-sentences may be ill-typed, all sentences in the image of the translation (and therefore all sentences relevant for the satisfaction condition) are well-typed.

Proposition 2. *For finite signatures, the model translation components of the comorphism from $SulCFOL^=$ to SoftFOL are surjective.*

Proof. Since we have restricted ourselves to $SulCFOL^=$, in a finite signature, each connected component of the subsort graph has a top element. A CASL model can be turned into a *SoftFOL* model as follows: take the disjoint union of all carriers of top sorts to be the carrier of the *SoftFOL* model. Predicates of same name and arity of the CASL model are united into a single predicate in the *SoftFOL* model. Similarly with operations, where operations need to be extended to the whole carrier of the *SoftFOL* model in an arbitrary way.   □

Corollary 3. *We can use the borrowing technique along the comorphism from CASL to SoftFOL. That is, CASL proof goals can be translated to SoftFOL proof goals in a sound and complete way.*

Proof. We assume that proof goals live over finite signatures and hence can apply Prop. 2. It is well known (see e.g. [4]) that surjectivity of the model translation leads to the property

$$\Gamma \models_\Sigma \varphi \text{ iff } Ax(\Phi(\Sigma)) \cup \alpha_\Sigma[\Gamma] \models_{Sign(\Phi(\Sigma))} \alpha_\Sigma(\varphi).$$

$\square$

5.2 Generating DFG Format from *SoftFOL*

A *SoftFOL* theory is transformed into a list of symbols, a list of declarations, and a list of axioms. The symbol list distinguishes sorts, predicates and functions and each predicate and function symbol is paired with its arity. Each subsort relation is transformed into a subsort declaration. The profiles of predicates and functions are also transformed into corresponding declarations. The sort generation axioms are turned into declarations of generated sorts (and to free generatedness, if the corresponding axioms are present). Note that declarations are treated by SPASS as special sentences.

5.3 Generating TPTP Format from *SoftFOL*

Since the TPTP language has no notion of a signature, only type information given in the signature is taken into account. A *SoftFOL* theory is transformed into a list of sentences named *declarationX* where X is a natural number. Each subsort relation and each function type is turned into a corresponding implication treating sorts as unary predicates. Typing information of predicates is not used. Concerning sentences, each variable list in a quantification that has type information is turned into an antecedent of an implication with the original quantified formula as consequent. The sorts are used like unary predicates. All other sentence constructs in *SoftFOL* have corresponding constructs in TPTP.

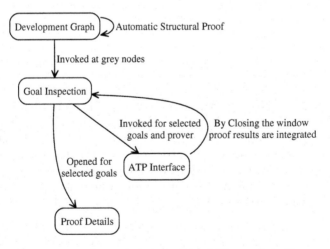

Fig. 2. Proof work-flow used for ATP proofs in HETS

6   Integration of ATP Reasoning into HETS

The logics and comorphisms described in the previous sections are integrated into the Heterogeneous Tool Set HETS, which already offers static analysis of CASL specifications and libraries and an automatic proof system at the structuring level of CASL. The Graphical User Interface (GUI) initially presents an analyzed and structurally proved development graph and offers to open the Goal Inspection GUI for the selection of goals and theorem provers at nodes with proof obligations, which are colored grey (see Fig. 2 for the proof work-flow and see Fig. 3 for a development graph).

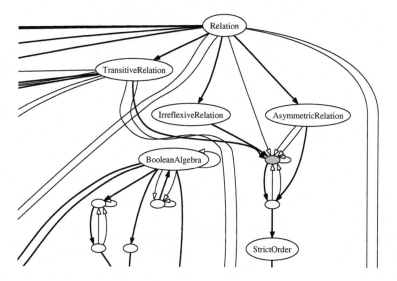

Fig. 3. Extract of the RELATIONSANDORDERS development graph

Figure 4 shows the interface for inspecting goals and discharging goals to different provers. The list on the left shows all goal names prefixed with the proof status in square brackets (see Tab. 2).

The two lists at the bottom of the window allow the detailed selection of axioms and proved theorems of this theory (see Sect. 7). By pressing the 'Prove' button the ATP interface of the selected prover is opened with the selected goals. The shortest (composed) comorphism is used for the translation into a prover

Table 2. Possible proof statuses

[+]	proved goal
[-]	disproved goal
[×]	proved goal revealing an inconsistent theory
[ ]	open goal

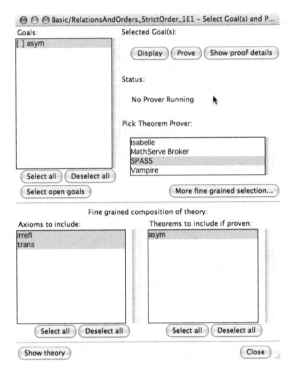

Fig. 4. HETS Goal Inspection Interface

supported logic. The button 'More fine grained selection...' allows the user to pick a (composed) comorphism in a separate window from where the prover interface is launched.

Currently the ATP systems SPASS and Vampire are connected to HETS and the MathServe broker, which classifies the theory and chooses a suitable ATP system for the proof attempt. The ATP interface for *SoftFOL* based provers is implemented generically, such that each interface has the same layout. Figure 5 shows a screen shot of the ATP interface instantiated for SPASS. The Vampire interface looks the same except for the window title and the broker interface has no extra options fields.

The ATP interface offers functions to call and inspect the selected goal in the upper part and the indicators in the goal list are the same as in the goal inspection interface (see Tab. 2). The batch mode tries to prove each open goal without interaction. The details shown for each goal in the ATP interface are specific to the different provers, while the proof tree integrated into the development graph is always in the TSTP format which is the unified TPTP solution format.

After closing the ATP interface the goal inspection interface (see Fig. 4) gets the proof results and offers a unified view at the structural level. Figure 6 shows the proof details for some goal proved with SPASS where the tactic script and the proof tree are hidden initially and the underlined labels are used to toggle the information hiding. The (composed) comorphism used for the translation into

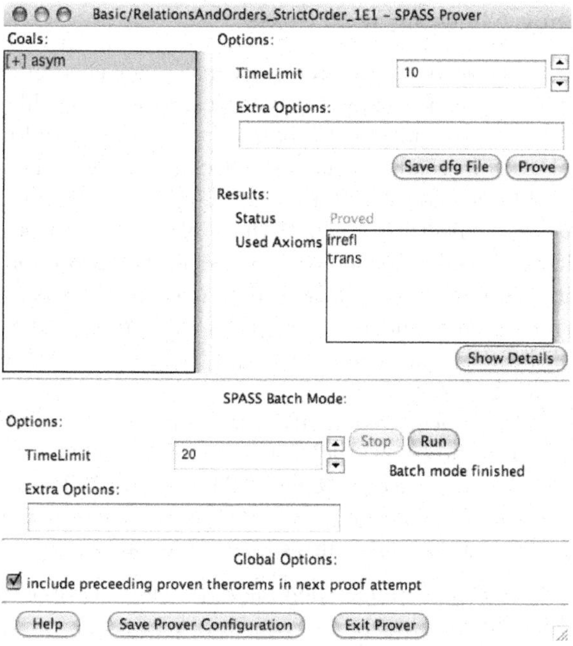

Fig. 5. ATP Interface of the SPASS prover

SoftFOL is shown as well. This example is taken from the specification STRIC-TORDER in the library BASIC/RELATIONSANDORDERS.

6.1 Connection of the ATPs

While SPASS is supposed to be installed and run locally with HETS on the same computer, MathServe is running on a dedicated central server which offers its reasoning services through HTTP and SOAP. The communication with SPASS is done via a standard Unix pipe and for MathServe all encoding, communication, and parsing of XML, SOAP, and HTTP is done through the HETS binary.

```
asym
    Com: SuleCFOL2SoftFOL : CASL -> SoftFOL
        Status: Proved
        Used axioms: "irrefl", "trans"
        Prover: SPASS
        Tactic script
        Time limit: 20
        Extra options: ["-DocProof"]

    Proof tree
```

Fig. 6. Proof details for goal asym

6.2 Consistency Checking with SPASS

Because SPASS can also disprove theorems, a consistency checker based on SPASS seemed appropriate. Because consistency checking of large libraries is time-consuming and hence more suitable for batch processing, we did not integrate it into the GUI of HETS. Instead, consistency checking is invoked outside of HETS after generating DFG problem files for each CASL specification in a library with logical atom *false* as conjecture. Each of these files is run through SPASS with time limit t (initially, t is 500 seconds). If SPASS finds a proof for false from the current theory it is inconsistent. If false is disproved SPASS has found a completion for the current theory and it is logged as consistent. But, often the time limit is exceeded, because the search space for the given theory is too large. In this case, the theory is t-consistent.

We have taken CASL specifications from the repository available under www. cofi.info/Libraries and tested their 500-second-consistency. Actually, this already revealed a number of inconsistencies. Passing to 1500-second-consistency increased the number of inconsistencies, while at 5000 seconds, a kind of saturation could be observed. Inconsistencies have been caused by

- missing definedness conditions when using partial functions within axioms, that is, formulas of the shape

$$\forall x : s \,.\, f(x) = t,$$

which force f to be total. The correct form is

$$\forall x : s \,.\, def\ x \Rightarrow f(x) = t$$

- missing side-conditions ensuring the well-definedness of operations on non-freely generated datatypes (such as sets), that is, formulas like

$$leftSummand(x \cup y) = x$$

- erroneously declaring a partial function as total,
- oversight of CASL's non-empty carrier assumption when defining subsorts, that is, declarations of form

$$\textbf{sort}\ s = \{x : s' \bullet \varphi(x)\}$$

with the possibility that there is no x with $\varphi(x)$ in some circumstances.

It should be noted that consistency checking with SPASS only concerns the first-order fragment of CASL specifications. That is, even when SPASS proves a theory to be consistent, using saturation, there can still be inconsistencies due to the presence of sort generation constraints.

6.3 Induction

SPASS uses sort generation constraints only for obtaining efficient rewriting strategies, and the other provers do not use these at all. That is, all provers that we have considered in this work are pure first-order provers. Still, we have realised a strategy how to perform induction proofs using first order provers: Along with the first-order goals sent to the prover, HETS takes the induction principles corresponding to sort generation constraints and instantiates them for the given proof goals. This already has been used for proving a number of theorems about inductive datatypes. Of course, with this method, one cannot expect the prover to find intermediate lemmas and prove them with induction – rather, the user has to provide the lemmas as proof goals along with the specification. Still, the method has turned out to be quite efficient: the fact that $reverse(reverse(L)) = L$ for any list L has been proved with just one lemma, which is much less than in the standard proof in Isabelle.

For induction proofs, it is crucial to carefully select the axioms that are fed into the first-order prover, otherwise it is very easy to run into a time-out (see also the next sections).

7   Used Optimisations

This section discusses some optimisations that we have implemented for the comorphism $SulCFOL^=$ to $SoftFOL$ and the encodings into the DFG and TPTP syntaxes.

Most of the optimisations are needed to shorten the search space of the proof and have been discussed with the developers of SPASS: Christoph Weidenbach, Thomas Hillenbrand, and Dalibor Topić. The central feature shortening the search space is the selection of axioms and proved goals that are included in the theory sent to the ATP system (see Fig. 4).

The more specialized optimizations are the following ones. Single sorted theories are translated into untyped FOL theories, because the sort provides no additional information in a single sorted theory; if the single sort is generated, it is still kept in the theory. If a sort is generated by constant constructors only, a sentence stating the exhaustive generation is introduced. Binary predicates equivalent to the built-in equality predicate are removed from the signature; the definitions of such binary predicates are removed from the sentences and the application of such binary predicates is substituted with the built-in equality in the other sentences. Unique existential quantification is coded out into a conjunction of two formulas: (1) there exists an element that satisfies the quantified formula and (2) all elements fulfilling the formula are equal to the element of the first conjunct. The conditional term construct is coded out in a standard way as a conjunction of two implications [5].

Before the DFG and TPTP format generation those signature elements, which are not used in the axioms and conjecture sent to the ATP system, are removed, except for sorts. This removes the declaration sentences of unused symbols from the theories. Sentences related to the definedness encoding of partial functions are not considered for finding used symbols. All sort injection functions, which are not used as constructors for generated sorts, are removed from the $SoftFOL$ signature.

8 Conclusion and Future Work

We have connected SPASS and several other automated theorem provers to CASL. This has been done via an institution comorphism to an intermediate logic *Soft-FOL* (softly typed first logic). Both the comorphism and the prover interfaces are integrated into the Heterogeneous Tool Set. The resulting tool has been used to verify a number of properties of CASL specifications, in particular for the complete verification of the composition table of the qualitative spatial calculi RCC5 and RCC8 based on Bennett's first-order axiomatization of connectedness [2]. The latter verification goal involved 95 theorems and turned out to be much too tedious to be carried out with the interactive prover Isabelle [12], which previously was the only prover for CASL available in a stage beyond initial prototypes. With SPASS, now a higher degree of automation is available.

The possibility of using the *SoftFOL* coding for checking consistency of CASL specifications turned out to be extremely useful – several typical specification errors could be found in a number of specifications. It is therefore advisable to check specifications in this way before proceeding e.g. to further proofs or refinements.

Actually, the SPASS developers often have the opinion that certain specification styles are just wrong, because they lead to theories that are difficult to handle for SPASS. We have answered this by implementing several optimizations according to feedback from the SPASS developers about unsuccessful proof attempts. We think that it is more the responsibility of tools to transform specifications that are "bad" for provers to "good" ones, and not the responsibility of the specifier – especially since the latter has to consider other goals such as clarity and validity of specifications.

To obtain further optimization, we have considered analyzing the SPASS output of unsuccessful proof attempts in order to obtain hints which axioms to exclude from the next proof attempt because they lead the prover into infinite loops. However, we think that such an analysis should be done by SPASS itself, since it can lead to a more fined grained penalty system that only gradually removes axioms from the list of used input formulas. The SPASS developers are somewhat reluctant to follow this path, as they claim that it is difficult to design general strategies detecting loops, and they point to the fact that the general problem is undecidable. We nevertheless think that it is worth trying to obtain at least some partial information along these lines and to use it for loop avoidance.

Currently, proofs are only inspected to obtain the list of used axioms, information which is essential for efficient change management. A future extension of the tool described here will translate the proof trees returned by the various provers into a format that is more easily readable by the CASL specifier.

Acknowledgements

We thank Christoph Weidenbach, Thomas Hillenbrand, and Dalibor Topić for SPASS and fruitful discussions, Christian Maeder for weekly consistency checks,

Jürgen Zimmer for MathServe and fruitful discussions, Stefan Wölfl for the RCC8 case study, Hennes Maertins and Lutz Schröder for the *reverse* example, Rene Wagner and Rainer Grabbe for help with the implementation, and Erwin R. Catesbeina for pointing out quite a number of inconsistencies.

References

1. Common framework initiative for algebraic specification and development. www.cofi.info.
2. B. Bennett. *Logical Representations for Automated Reasoning about Spatial Relationships*. PhD thesis, 1997.
3. M. Bidoit and P. D. Mosses. CASL *User Manual*, volume 2900 of *LNCS*. Springer Verlag; Berlin, 2004. With chapters by Till Mossakowski, Donald Sannella, and Andrzej Tarlecki.
4. M. Cerioli and J. Meseguer. May I borrow your logic? (transporting logical structures along maps). *Theoretical Comput. Sci.*, 173:311–347, 1997.
5. CoFI (The Common Framework Initiative). CASL *Reference Manual*, volume 2960 of *LNCS*. Springer Verlag; Berlin, 2004.
6. J. Goguen and G. Rosu. Institution morphisms. *Formal aspects of computing*, 13:274–307, 2002.
7. J. A. Goguen and R. M. Burstall. Institutions: Abstract model theory for specification and programming. *Journal of the Association for Computing Machinery*, 39:95–146, 1992. Predecessor in: LNCS 164, 221–256, 1984.
8. J. A. Goguen and J. Meseguer. Order-sorted algebra I: equational deduction for multiple inheritance, overloading, exceptions and partial operations. *Theoretical Comput. Sci.*, 105:217–273, 1992.
9. J. Meseguer. Membership algebra as a logical framework for equational specification. In F. Parisi Presicce, editor, *Recent trends in algebraic development techniques. Proc. 12th International Workshop*, volume 1376 of *Lecture Notes in Computer Science*, pages 18–61. Springer, 1998.
10. T. Mossakowski. Relating CASL with other specification languages: The institution level. *Theoretical Comput. Sci.*, 286:367–475, 2002.
11. T. Mossakowski, C. Maeder, and K. Lüttich. The Heterogeneous Tool Set. Available at www.tzi.de/cofi/hets, University of Bremen.
12. T. Nipkow, L. C. Paulson, and M. Wenzel. *Isabelle/HOL — A Proof Assistant for Higher-Order Logic*, volume 2283 of *LNCS*. Springer Verlag; Berlin, 2002.
13. G. Sutcliffe and C. Suttner. The TPTP Problem Library: CNF Release v1.2.1. *Journal of Automated Reasoning*, 21(2):177–203, 1998.
14. G. Sutcliffe and C. Suttner. The State of CASC. *AI Communications*, 19(1):35–48, 2006.
15. C. Weidenbach, U. Brahm, T. Hillenbrand, E. Keen, C. Theobalt, and D. Topic. SPASS version 2.0. In A. Voronkov, editor, *Automated Deduction – CADE-18*, volume 2392 of *LNCS*, pages 275–279. Springer Verlag; Berlin, 2002.
16. J. Zimmer and S. Autexier. The MathServe System for Semantic Web Reasoning Services. In U. Furbach and N. Shankar, editors, *Proceedings of the third International Joint Conference on Automated Reasoning*, volume 4130 of *LNCS*, pages 140–144. Springer Verlag; Berlin, 2006.

Structured CSP –
A Process Algebra as an Institution[*]

Till Mossakowski[1] and Markus Roggenbach[2]

[1] DFKI Lab Bremen and University of Bremen, Germany
till@tzi.de
[2] University of Wales Swansea, United Kingdom
M.Roggenbach@Swan.ac.uk

Abstract. We introduce two institutions for the process algebra CSP, one for the traces model, and one for the stable failures model. The construction is generic and should be easily instantiated with further models. As a consequence, we can use structured specification constructs like renaming, hiding and parameterisation (that have been introduced over an arbitrary institution) also for CSP. With a small example we demonstrate that structuring indeed makes sense for CSP.

1 Introduction

Among the various frameworks for the description and modelling of reactive systems, process algebra plays a prominent role. Here, the process algebra CSP [13, 18] has successfully been applied in various areas, ranging from train control systems [7] over software for the international space station [6] to the verification of security protocols [19].

In this paper we extend the process algebra CSP by a 'module concept' that allows us to build complex specifications out of simpler ones. To this end, we re-use typical structuring mechanisms from algebraic specification as they are realised, e.g., in the algebraic specification language CASL [8, 4]. This approach leads to a new specification paradigm for reactive systems: our framework offers also the loose specification of CSP processes, where the structured free construct applied to a basic specification yields the usual fixed point construction by Tarski's theorem.

On the theoretical side our approach requires us to formulate the process algebra CSP as an institution [12] — the latter notion captures the essence of a logical system and allows for logic-independent structuring languages. We show that various CSP models[1] fit into this setting. The practical outcome is a flexible module concept. We demonstrate through some examples that these structuring

---

[*] This work has been supported by EPSRC under the grant EP/D037212/1 and by the German DFG under grant KR 1191/5-2.
[1] i.e. the combination of process syntax, semantic domain, semantic clauses, and a fixed-point theory in order to deal with recursion.

J.L. Fiadeiro and P.-Y. Schobbens (Eds.): WADT 2006, LNCS 4409, pp. 92–110, 2007.

mechanisms (e.g. extension, union, renaming, parametrisation) are suitable for CSP. Furthermore, formulating a process algebra as an institution links two hitherto unrelated worlds.

The paper is organised as follows: Sect. 2 discusses what a CSP signature might be. Then we describe in a generic way how to build a CSP institution. It turns out that many properties can already be proven in the generic setting. Sections 4 and 5 instantiate the generic institution with the traces model and the stable failures model, resp. Having now institutions available, we discuss how to obtain the full range of structuring mechanisms in spite of the missing pushouts of our signature category. In Sect. 7 we make structured specifications available to CSP and illustrate this with a classical example of process algebra. Sect. 8 discusses some related work and concludes the paper.

2 What Is an Appropriate Notion of a Signature Morphism?

When analysing CSP specifications, it becomes clear that there are two types of symbols that change from specification to specification: communications and process names. Pairs consisting of an alphabet A of communication symbols and of process names N (together with some type information) will eventually be the objects of our category CSPSIG of CSP signatures, see Sect 3.1 below. The notion of a signature morphism, however, is not as easy to determine. An institution captures how truth can be preserved under change of symbols. In this sense, we want to come up with a notion of a signature morphism that is as liberal as possible but still respects fundamental CSP properties. In this section we discuss why this requires to restrict alphabet translations to injective functions.

The process algebra CSP itself offers an operator that changes the communications of a process P, namely *functional renaming*[2] $f[P]$. Here, $f : A \to ?\ A$ is a (partial) function such that $dom(f)$ includes all communications occurring in P. The CSP literature, see e.g. [18], classifies functional renaming as follows: (1) Functional renaming with an injective function f preserves all process properties. (2) Functional renaming with a non-injective function f is mainly used for process abstraction. Non-injective renaming can introduce unbounded non-determinism[3], and thus change fundamental process properties.

As a process algebra, CSP exhibits a number of fundamental algebraic laws. Among these the so-called step laws of CSP, take for example the following law ⟨□-**step**⟩,

---

[2] Note that the so-called relational renaming, which is included in our CSP dialect, subsumes functional renaming.

[3] Take for example $f[?n : \mathbf{N} \to (-n) \to Skip] = 0 \to \sqcap \{(-n) \to Skip \mid n \in \mathbf{N}\}$, where $f(z) = 0$, if $z \geq 0$, and $f(z) = z$, if $z < 0$. As functional renaming can be expressed in terms of relational renaming, the process on the left-hand side is part of our CSP dialect. The process on the right-hand side, however, does not belong to our CSP dialect, as we restrict the internal choice operator to be binary only.

$$(?x : A \to P) \,\square\, (?y : B \to Q)$$
$$= ?x : A \cup B \to if\ x \in A \cap B\ then\ (P \sqcap Q)\ else\ (if\ x \in A\ then\ P\ else\ Q)$$

are of a special significance: The step laws do not only hold in all the main CSP models, including the traces model $\mathcal{T}$, the failures/divergences model $\mathcal{N}$, and the stable-failures model $\mathcal{F}$. They are also essential for the definition of complete axiomatic semantics for CSP, see [18, 14]. The CSP step laws show that e.g. the behaviour of external choice $\square$, alphabetised parallel $\|[X]\|$ and hiding \ crucially depends on the equality relation in the alphabet of communications. We demonstrate this here for the external choice operator $\square$:

– Assume $a \neq b$. Then

$$(?x : \{a\} \to P) \,\square\, (?y : \{b\} \to Q)$$
$$= ?x : \{a, b\} \to if\ x \in \{a\} \cap \{b\}\ then\ (P \sqcap Q)\ else\ (if\ x \in \{a\}\ then\ P\ else\ Q)$$
$$= ?x : \{a, b\} \to if\ x \in \{a\}\ then\ P\ else\ Q$$

– Mapping a and b with a non-injective function f to the same element c has the effect:

$$f[(?x : \{a\} \to P) \,\square\, (?y : \{b\} \to Q)]$$
$$= ((?x : \{c\} \to f[P]) \,\square\, (?y : \{c\} \to f[Q]))$$
$$= ?x : \{c\} \to if\ x \in \{c\} \cap \{c\}\ then\ (f[P] \sqcap f[Q])\ else$$
$$\qquad (if\ x \in \{c\}\ then\ f[P]\ else\ f[Q])$$
$$= ?x : \{c\} \to (f[P] \sqcap f[Q])$$

I.e. before the translation, the environment controls which one of the two processes P and Q is executed - after the translation this control has been lost: The process makes an internal choice between $f[P]$ and $f[Q]$. Similar examples can be extracted from the step laws for external choice $\square$, alphabetised parallel $\|[X]\|$ and hiding \.

Summarised: Non-injective renaming can fundamentally change the behaviour of processes. One reason for this is that alphabets of communications play two roles in CSP: They are constituents of both (i) the process syntax and (ii) the semantic domain. This causes problems with non-injective functions as signature morphisms: syntax is translated covariantly while semantics is translated contravariantly.

3   The CSP Institution – General Layout

Institutions have been introduced by Goguen and Burstall [12] to capture the notion of logical system and abstract away from the details of signatures, sentences, models and satisfaction. We briefly recall the notion here.

Let $\mathcal{CAT}$ be the category of categories and functors.[4]

---

[4] Strictly speaking, $\mathcal{CAT}$ is not a category but only a so-called quasi-category, which is a category that lives in a higher set-theoretic universe.

Definition 1. *An* institution $I = (\mathbf{Sign}, \mathbf{Sen}, \mathbf{Mod}, \models)$ *consists of*

- *a category* **Sign** *of signatures,*
- *a functor* **Sen**: **Sign** $\longrightarrow$ **Set** *giving, for each signature* Σ, *the set of sentences* **Sen**(Σ), *and for each signature morphism* $\sigma: \Sigma \longrightarrow \Sigma'$, *the sentence translation map* **Sen**(σ): **Sen**$(\Sigma) \longrightarrow$ **Sen**(Σ'), *where often* **Sen**$(\sigma)(\varphi)$ *is written as* $\sigma(\varphi)$,
- *a functor* **Mod**: **Sign**$^{op} \longrightarrow \mathcal{CAT}$ *giving, for each signature* Σ, *the category of models* **Mod**(Σ), *and for each signature morphism* $\sigma: \Sigma \longrightarrow \Sigma'$, *the reduct functor* **Mod**(σ): **Mod**$(\Sigma') \longrightarrow$ **Mod**(Σ), *where often* **Mod**$(\sigma)(M')$ *is written as* $M'|_\sigma$,
- *a satisfaction relation* $\models_\Sigma \subseteq |$ **Mod**$(\Sigma) | \times$ **Sen**(Σ) *for each* $\Sigma \in |$ **Sign** $|$,

such that for each $\sigma: \Sigma \longrightarrow \Sigma'$ *in* **Sign** *the following* satisfaction condition *holds:*

$$M' \models_{\Sigma'} \sigma(\varphi) \Leftrightarrow M'|_\sigma \models_\Sigma \varphi$$

for each $M' \in |$ **Mod**$(\Sigma') |$ *and* $\varphi \in$ **Sen**(Σ).

We first discuss the general layout of the CSP institution independently of a concrete CSP model.

3.1   The Category CSPSIG of CSP Signatures

An *object* in the category CSPSIG is a pair (A, N) where

- A is an alphabet of communications and
- $N = (\bar{N}, sort, param)$ collects information on process names; $\bar{N}$ is a set of process names, where each $n \in \bar{N}$ has
  - a parameter type $param(n) = \langle X_1, \ldots, X_k \rangle$, $X_i \subseteq A$ for $1 \leq i \leq k$, $k \geq 0$. A process name without parameters has the empty sequence $\langle \rangle$ as its parameter type.
  - a type $sort(n) = X \subseteq A$, which collects all communications in which the process n can possibly engage in.

By abuse of notation, we will write $n \in N$ instead of $n \in \bar{N}$ and $(a_1, \ldots, a_k) \in param(n)$ instead of $(a_1, \ldots, a_k) \in X_1 \times \ldots \times X_k$, where $param(n) = \langle X_1, \ldots, X_k \rangle$.

A *morphism* $\sigma = (\alpha, \nu) : (A, N) \rightarrow (A', N')$ in the category CSPSIG consists of two maps

- $\alpha : A \rightarrow A'$, an injective translation of communications, and
- $\nu : N \rightarrow N'$, a translation of process names, which has the following two properties:
  - $param'(\nu(n)) = \alpha(param(n))$: preservation of parameter types, where $\alpha(param(n))$ denotes the extension of α to sequences of sets.
  - $sort'(\nu(n)) \subseteq \alpha(sort(n))$: non-expansion of types, i.e. the translated process $\nu(n)$ is restricted to those events which are obtained by translation of its type $sort(n)$.

The non-expansion of types principle is crucial for ensuring the satisfaction condition of the CSP institution below. It ensures that the semantics of a process is frozen when translated to a larger context, i.e. even when moving to a larger alphabet, up to renaming, models for "old" names may only use "old" alphabet letters. This corresponds to a *black-box* view on processes that are imported from other specification modules.

As usual, the *composition of morphisms* $\sigma = (\alpha, \nu) : (A, N) \rightarrow (A', N')$ and $\sigma' = (\alpha', \nu') : (A', N') \rightarrow (A'', N'')$ is defined as $\sigma' \circ \sigma := (\alpha' \circ \alpha, \ \nu' \circ \nu)$.

3.2 Sentences

Given A : alphabet of communications
N : set of process names
Z : variable system over A
$\mathcal{L}(A, N, Z)$: logic

we define

$P, Q ::=$	$n(z_1, \ldots, z_k)$	%% (possibly parametrised) process name
	$Skip$	%% successfully terminating process
	$Stop$	%% deadlock process
	$a \rightarrow P$	%% action prefix with a communication
	$y \rightarrow P$	%% action prefix with a variable
	$?x : X \rightarrow P$	%% prefix choice
	$P \,\square\, Q$	%% external choice
	$P \,\sqcap\, Q$	%% internal choice
	$if\ \varphi\ then\ P\ else\ Q$	%% conditional
	$P \,[\![\, X \,]\!]\, Q$	%% generalized parallel
	$P \setminus X$	%% hiding
	$P[[r]]$	%% relational renaming
	$P \,\fatsemi\, Q$	%% sequential composition

where

$n \in N$, $param(n) = \langle X_1, \ldots, X_k \rangle$ for some $k \in \mathbf{N}$, and $z_i \in (\bigcup_{Y \subseteq X_i} Z_Y) \cup X_i$ for $1 \leq i \leq k$; $a \in A$; $y \in Z$; $x \in Z_X$; $X \subseteq A$; $\varphi \in \mathcal{L}(A, N, Z)$ is a formula; and $r \subseteq A \times A$

Fig. 1. CSP syntax

Relative to an alphabet of communications A we define a *variable system* $Z = (Z_X)_{X \in \mathcal{P}(A)}$ to be a pairwise disjoint family of variables, where subsets $X \subseteq A$ of the alphabet A are the indices.

The standard CSP literature does not reflect what kind of logic $\mathcal{L}(A, N, Z)$ is plugged into the language. A logic that is quite simple but covers those formulae usually occurring in process examples is given in Fig. 2. We record some properties of formulae:

L1 There is a substitution operator $[b/y]$ defined in an obvious way on formulae. Substitution has the following property: If $\varphi \in \mathcal{L}(A, N, Z)$, $y : Y \in Z$, and $b \in Y$, for some $Y \subseteq A$, then $\varphi[b/y] \in \mathcal{L}(A, N, Z\backslash\{y : Y\})$.

CSP *terms*, see Fig. 1 for the underlying grammar, are formed relatively to a signature (A, N), a variable system Z over A, and a logic $\mathcal{L}(A, N, Z)$. Additional CSP operators can be encoded as syntactic sugar, including the synchronous parallel operator $P \parallel Q := P \parallel A \parallel Q$ and the interleaving operator $P \parallel\!\parallel Q := P \parallel \emptyset \parallel Q$.

For the purpose of turning CSP into an institution, the use of variables needs to be made more precise. Given a system of *global variables* G and a system of *local variables* L, which are disjoint, we define the system of *all variables* $Z := G \cup L$. We define the set of process terms $T_{(A,N)}(G, L)$ over a signature (A, N) to be the least set satisfying the following rules:

- $n(z_1, \ldots, z_k) \in T_{(A,N)}(G, L)$ if $n \in N$, $param(n) = \langle X_1, \ldots, X_k \rangle$ for some $k \in \mathbf{N}$, and $z_i \in (\bigcup_{Y \subseteq X_i} Z_Y) \cup X_i$ for $1 \leq i \leq k$;
- $Skip$, $Stop \in T_{(A,N)}(G, L)$.
- $a \rightarrow P \in T_{(A,N)}(G, L)$ if $a \in A$ and $P \in T_{(A,N)}(G, L)$
- $x \rightarrow P \in T_{(A,N)}(G, L)$ if $x \in G \cup L$ and $P \in T_{(A,N)}(G, L)$
- $?x : X \rightarrow P \in T_{(A,N)}(G, L)$ if $P \in T_{(A,N)}(G, L \cup \{x : X\})$.
- $P \,\square\, Q$, $P \sqcap Q \in T_{(A,N)}(G, L)$ if $P, Q \in T_{(A,N)}(G, L)$.
- *if* φ *then* P *else* $Q \in T_{(A,N)}(G, L)$ if $P, Q \in T_{(A,N)}(G, L)$ and $\varphi \in \mathcal{L}(A, N, Z)$.
- $P \parallel X \parallel Q \in T_{(A,N)}(G, L)$ if $P, Q \in T_{(A,N)}(G, L)$ and $X \subseteq A$.
- $P \setminus X \in T_{(A,N)}(G, L)$, if $P \in T_{(A,N)}(G, L)$ and $X \subseteq A$.
- $P[[r]] \in T_{(A,N)}(G, L)$ if $P \in T_{(A,N)}(G, L)$ and $r \subseteq A \times A$.
- $P \,\mathring{,}\, Q \in T_{(A,N)}(G, L)$ if $P \in T_{(A,N)}(G, L)$ and $Q \in T_{(A,N)}(G, \emptyset)$.

The set of global variables remains constant in all rules; local variables are effected in the rules for prefix choice and sequential composition: prefix choice adds a new local variable; sequential composition deletes all local variables.

The CSP semantics deals with variables using substitution on the syntax level. Here, $P[b/y]$ denotes the process P in which every free occurrence of the variable

---

> *Formulae in $\mathcal{L}(A, N, Z)$:*
> $t_1 = t_2$ t_1, t_2 terms over (N, A) and Z
> $t \in X$ t a term over (N, A) and Z; $X \subseteq A$
>
> *Terms over (N, A) and Z:*
> a $a \in A$ (alphabet symbol)
> x $x \in Z$ (variable)

Fig. 2. A simple logic for formulae occurring in CSP processes

$$
\begin{aligned}
n(z_1, \ldots, z_k)[a/y] &= n(y_1, \ldots, y_k) \text{ with } y_i = \begin{cases} a & \text{if } z_i = y \\ z_i & \text{otherwise.} \end{cases} \\
Skip[b/y] &= Skip \\
Stop[b/y] &= Stop \\
(a \to P)[b/y] &= a \to P[b/y] \\
(x \to P)[b/y] &= \begin{cases} b \to P[b/y] \; ; & x = y \\ x \to P[b/y] \; ; & x \neq y \end{cases} \\
(?x : X \to P)[b/y] &= \begin{cases} ?x : X \to P & ; \; x = y \\ ?x : X \to P[b/y] & ; \; x \neq y \end{cases} \\
(P \,\square\, Q)[b/y] &= P[b/y] \,\square\, Q[b/y] \\
(P \,\sqcap\, Q)[b/y] &= P[b/y] \,\sqcap\, Q[b/y] \\
(if \; \varphi \; then \; P \; else \; Q)[b/y] &= if \; \varphi[b/y] \; then \; P[b/y] \; else \; Q[b/y] \\
(P \,[\![\, X \,]\!]\, Q)[b/y] &= P[b/y] \,[\![\, X \,]\!]\, Q[b/y] \\
(P \setminus X)[b/y] &= P[b/y] \setminus X \\
(P[[r]])[b/y] &= (P[b/y])[[r]] \\
(P \,\fatsemi\, Q)[b/y] &= P[b/y] \,\fatsemi\, Q[b/y]
\end{aligned}
$$

Fig. 3. Substitution

$y : Y$ is replaced by a communication $b \in Y$. Fig. 3 gives the formal definition[5]. We write $P[a_1/x_1, a_2/x_2, \ldots, a_n/x_n]$ for $(\ldots ((P[a_1/x_1])[a_2/x_2]) \ldots)[a_n/x_n]$.

A *process definition* over a signature (A, N) is an equation

$$ p(x_1, \ldots, x_k) = P $$

where $p \in N$, the x_i are variables with $x_i : X_i$, where X_i is the *i-th* component of $param(p)$, and P is a term. A process definition is a *sentence* if $P \in T_{(sort(p), N)}(\{x_1 : X_1, \ldots, x_k : X_k\}, \emptyset)$.

3.3   Translation Along a Signature Morphism

Let $\sigma = (\alpha, \nu) : (A, N) \to (A', N')$ be a signature morphism. Given a variable system $Z = (Z_X)_{X \in \mathcal{P}(A)}$ over (A, N) we obtain a variable system $\sigma(Z)$ over (A', N') by $\sigma(Z)_{X'} := \bigcup_{\alpha(X) = X'} Z_X$. For an individual variable $x : X$ this translation yields $\sigma(x : X) = \alpha(x : X) = x : \alpha(X)$. For the *translation of formulae* we require:

L2 $\mathcal{L}$ has a formula translation of the type $\sigma : \mathcal{L}(A, N, Z) \to \mathcal{L}(A', N', \sigma(Z))$ with the following property: given a formula $\varphi \in \mathcal{L}(A, N, Z)$, then $\sigma(\varphi) \in \mathcal{L}(\alpha(A), N', \sigma(Z))$.

L3 Formula translation composes, i.e., for all signature morphisms $\sigma = (\alpha, \nu) : (A, N) \to (A', N')$, $\sigma' = (\alpha', \nu') : (A', N') \to (A'', N'')$, and $\varphi \in \mathcal{L}(A, N, Z)$ holds: $(\sigma' \circ \sigma)(\varphi) = \sigma'(\sigma(\varphi))$.

---

[5] The rule for prefix choice deals with free and bound variables. In the case of sequential composition only a substitution with a global variable can have an effect on the process Q.

Properties **L2** and **L3** are indeed satisfied by our simple logic given in Fig. 2.

Fig. 4 gives the rules for *term translation. Translation of process definitions* is defined as

$$\sigma(p(x_1, \ldots, x_k) = P) := \sigma(p(x_1, \ldots, x_k)) = \sigma(P).$$

The translation of process definitions composes.

$$
\begin{aligned}
\sigma(n(z_1, \ldots, z_k)) &:= \nu(n)(\alpha(z_1), \ldots, \alpha(z_k)) & \sigma(P \sqcap Q) &:= \sigma(P) \sqcap \sigma(Q) \\
\sigma(Stop) &:= Stop & \sigma(if\ \varphi\ then\ P\ else\ Q) &:= if\ \sigma(\varphi)\ then\ \sigma(P) \\
\sigma(Skip) &:= Skip & & \quad else\ \sigma(Q) \\
\sigma(a \rightarrow P) &:= \alpha(a) \rightarrow \sigma(P) & \sigma(P \,[\![X]\!]\, Q) &:= \sigma(P) \,[\![\alpha(X)]\!]\, \sigma(Q) \\
\sigma(x \rightarrow P) &:= x \rightarrow \sigma(P) & \sigma(P \setminus X) &:= \sigma(P) \setminus \alpha(X) \\
\sigma(?x : X \rightarrow P) &:= ?x : \alpha(X) \rightarrow \sigma(P) & \sigma(P[\![r]\!]) &:= \sigma(P)[\![\alpha(r)]\!] \\
\sigma(P \,\square\, Q) &:= \sigma(P) \,\square\, \sigma(Q) & \sigma(P \,\mathbin{\raise0.3ex\hbox{$\scriptstyle\circ$}}_{\,9}\, Q) &:= \sigma(P) \,\mathbin{\raise0.3ex\hbox{$\scriptstyle\circ$}}_{\,9}\, \sigma(Q) \\
& \text{where } \alpha(r) := \{(\alpha(x), \alpha(y)) \mid (x, y) \in r\}
\end{aligned}
$$

Fig. 4. Term translation

3.4   Models and Reducts

Let $\mathcal{D}(A)$ be a CSP domain constructed relatively to a set of communications A. Examples of $\mathcal{D}(A)$ are the domain $\mathcal{T}(A)$ of the CSP traces model, see Section 4, and the domain $\mathcal{F}(A)$ of the CSP stable failures model, see Section 5. A *model* M over a signature (A, N) assigns to each n and for all $a_1, \ldots, a_k \in param(n)$ a type correct element of the semantic domain $\mathcal{D}(A)$, i.e.

$$M(n(a_1, \ldots, a_k)) \in \mathcal{D}(sort(n)) \subseteq \mathcal{D}(A).$$

We define model categories to be partial orders, that is, there is a morphism between models M_1 and M_2, iff $M_1 \sqsubseteq M_2$. Here $\sqsubseteq$ is the pointwise extension of the partial order used in the denotational CSP semantics for the chosen domain $\mathcal{D}$; see the individual domains for the concrete choice of the partial order.

Given an injective (total) alphabet translation $\alpha : A \rightarrow A'$ we define its partial inverse as

$$
\hat{\alpha} : \quad \begin{array}{l} A' \rightarrow? A \\ a' \mapsto \begin{cases} \hat{\alpha}(a) & ;\ if\ a \in A\ is\ such\ that\ \alpha(a) = a' \\ undefined & ;\ otherwise \end{cases} \end{array}
$$

Let $\hat{\alpha}_{\mathcal{D}} : \mathcal{D}(A') \rightarrow? \mathcal{D}(A)$ be the extension of $\hat{\alpha}$ to semantic domains – to be defined for any domain individually.

The *reduct* of a model M' along σ is defined as

$$M'|_\sigma (n(a_1, \ldots, a_k)) = \hat{\alpha}_{\mathcal{D}}(M'(\nu(n)(\alpha(a_1), \ldots, \alpha(a_k)))).$$

As for reducts it is clear that we work with domains, we usually omit the index and write just $\hat{\alpha}$. On the level of domains, we define the following **reduct condition** on α and $\hat{\alpha}$: $\forall X \subseteq A$: $\hat{\alpha}(\mathcal{D}(\alpha(X))) \subseteq \mathcal{D}(X)$.

Theorem 2 (Reducts are type correct). Let α and $\hat{\alpha}$ fulfil the reduct condition. Then reducts are type correct, i.e. $M'|_\sigma(n(a_1, \ldots, a_k)) \in \mathcal{D}(sort(n))$.

3.5   Satisfaction

Given a map $denotation : M \times P \to \mathcal{D}(A)$, which – given a model M – maps a closed process term $P \in T_{(A,N)}(\emptyset, \emptyset)$ to its denotation in $\mathcal{D}$, we define the satisfaction relation of our institution[6]:

$$M \models p(x_1, \ldots, x_k) = P$$
$$:\Leftrightarrow$$
$$\forall(a_1, \ldots, a_k) \in param(p).$$
$$denotation_M(p(a_1, \ldots a_k)) = denotation_M(P[a_1/x_1, \ldots, a_k/x_k])$$

Remark 3. We can replace the logic $\mathcal{L}(A, N, Z)$ by any other logic that comes with a satisfaction relation

$$\models \subseteq \text{CSPMOD}_\mathcal{D}(A, N) \times \mathcal{L}(A, N, \emptyset)$$

and satisfies laws **L1** to **L3** above, plus

L4 The logic fulfils a satisfaction condition, i.e., forall $\varphi \in \mathcal{L}(A, N, \emptyset)$ holds:

$$M'|_\sigma \models \varphi[a_1/x_1, \ldots, a_n/x_n] \Leftrightarrow M' \models \sigma(\varphi)[\alpha(a_1)/x_1, \ldots, \alpha(a_n)/x_n]$$

To be concise with the CSP semantics, which deals with variables using substitution on the syntax level, it is necessary to include here a (possibly empty) substitution, see the reduct property stated in Theorem 4 below.

The CSP models give interpretations to the process names. The formulae used in practical CSP examples usually only reason about data, not on processes. Thus, in the satisfaction condition above the notion of a model and its reduct will vanish in most logic instances.

If the chosen CSP model has the reduct property and the extension of α and $\hat{\alpha}$ are inverse functions on $\mathcal{D}(A)$ and $\mathcal{D}(\alpha(A))$, the satisfaction condition holds:

Theorem 4 (Satisfaction condition). Let $\sigma = (\alpha, \nu) : (A, N) \to (A', N')$ be a signature morphism. Let M' be a (A', N')-model over the domain $\mathcal{D}(A')$. Let the following **reduct property** hold:

$$denotation_{M'|_\sigma}(P[a_1/x_1, \ldots, a_n/x_n])$$
$$= \hat{\alpha}(denotation_{M'}(\sigma(P)[\alpha(a_1)/x_1, \ldots, \alpha(a_n)/x_n]))$$

for all $P \in T_{(A,N)}(\{x_1 : X_1, \ldots x_n : X_n\}, \emptyset)$, $a_i \in X_i \subseteq A$ for $1 \leq i \leq n$, $n \geq 0$. Let α and $\hat{\alpha}$ be inverses on $\mathcal{D}(A)$ and $\mathcal{D}(\alpha(A))$. Under these conditions, we have for all process definitions $p(x_1, \ldots, x_k) = P$ over (A, N):

$$M'|_\sigma \models p(x_1, \ldots, x_k) = P \Leftrightarrow M' \models \sigma(p(x_1, \ldots, x_k) = P).$$

---

[6] Here and in the following we use ':⇔' as an abbreviation for 'iff, by definition'.

4   The CSP Traces Model as an Institution

Given an alphabet A and an element $\checkmark \notin A$ (denoting successful termination) we define sets $A^{\checkmark} := A \cup \{\checkmark\}$ and $A^{*\checkmark} := A^* \cup \{t \frown \langle\checkmark\rangle \mid t \in A^*\}$. The domain $\mathcal{T}(A)$ of the traces model is the set of all subsets T of $A^{*\checkmark}$ for which the following healthiness condition holds:

T1 T is non-empty and prefix closed.

The domain $\mathcal{T}(A)$ gives rise to the notion of trace refinement $S \sqsubseteq_{\mathcal{T}} T :\Leftrightarrow T \subseteq S$. $(\mathcal{T}(A), \sqsubseteq_{\mathcal{T}})$ forms a complete lattice, with $A^{*\checkmark}$ as its bottom and $\{\langle\rangle\}$ as its top. *Morphisms* in the category $Mod_{\mathcal{T}}(A, N)$ are defined as:

$$M_1 \rightarrow M_2 :\Leftrightarrow$$
$$\forall n \in N : \forall a_1, \ldots, a_k \in param(n) : M_2(n(a_1, \ldots, a_k)) \sqsubseteq_{\mathcal{T}} M_1(n(a_1, \ldots, a_k))$$

$I(n(a_1, \ldots, a_k)) = \{\langle\rangle\}$, i.e. the model which maps all instantiated process names to the denotation of *Stop* is initial in $Mod_{\mathcal{T}}(A, N)$; $F(n(a_1, \ldots, a_k)) = A^{*\checkmark}$ is final in $Mod_{\mathcal{T}}(A, N)$.

Let $\sigma = (\alpha, \nu) : (A, N) \rightarrow (A', N')$ be a signature morphism. We extend the map α canonically to three maps $\alpha^{\checkmark}$, $\alpha^{*\checkmark}$ and $\alpha_{\mathcal{T}}^{*\checkmark}$ to include the termination symbol, to extend it to strings, and to let it apply to elements of the semantic domain, respectively. In the same way we can extend $\hat{\alpha}$, the partial inverse of α, to three maps $\hat{\alpha}^{\checkmark}$, $\hat{\alpha}^{*\checkmark}$ and $\hat{\alpha}_{\mathcal{T}}^{*\checkmark}$. With these notions, it holds that:

Theorem 5 (Reducts in the traces model are well-behaved)

1. Let $T' \in \mathcal{T}(A')$. Then $\hat{\alpha}(T') \in \mathcal{T}(A)$.
2. $\forall X \subseteq A :\ \hat{\alpha}(\mathcal{T}(\alpha(X))) \subseteq \mathcal{T}(X)$.

$$
\begin{aligned}
traces_M(n(a_1, \ldots, a_k)) &= M(n(a_1, \ldots, a_k)) \\
traces_M(Skip) &= \{\langle\rangle, \langle\checkmark\rangle\} \\
traces_M(Stop) &= \{\langle\rangle\} \\
traces_M(a \rightarrow P) &= \{\langle\rangle\} \cup \{\langle a\rangle \frown s \mid s \in traces_M(P)\} \\
traces_M(?\ x : X \rightarrow P) &= \{\langle\rangle\} \cup \{\langle a\rangle \frown s \mid s \in traces_M(P[a/x]), a \in X\} \\
traces_M(P \mathbin{\square} Q) &= traces_M(P) \cup traces_M(Q) \\
traces_M(P \sqcap Q) &= traces_M(P) \cup traces_M(Q) \\
traces_M(if\ \varphi\ then\ P\ else\ Q) &= if\ M \models \varphi\ then\ traces_M(P)\ else\ traces_M(Q) \\
traces_M(P \mathbin{[\![} X \mathbin{]\!]} Q) &= \bigcup\{t_1 \mathbin{[\![} X \mathbin{]\!]} t_2 \mid t_1 \in traces_M(P), t_2 \in traces_M(Q)\} \\
traces_M(P \setminus X) &= \{t \setminus X \mid t \in traces_M(P)\} \\
traces_M(P[[r]]) &= \{t \mid \exists\, t' \in traces_M(P).\ (t', t) \in r^*\} \\
traces_M(P \mathbin{\raise1pt\hbox{$\,{}_\circ^\circ\,$}} Q) &= (traces_M(P) \cap A^*) \cup \\
&\quad \{t_1 \frown t_2 \mid t_1 \frown \langle\checkmark\rangle \in traces_M(P), t_2 \in traces_M(Q)\}
\end{aligned}
$$

Fig. 5. Semantic clauses of the basic processes in the traces model $\mathcal{T}$

Fig. 5 gives the semantic clauses of the traces model, see [18] for the definition of the various operators on traces. Note that thanks to the rules imposed on the use of variables, there is no need to provide a denotation for a process term of the form $x \rightarrow P$: In the clause for prefix choice $?x : X \rightarrow P$, which is the only way to introduce a variable x, every free occurrence of x in the process P is syntactically substituted by a communication.

Lemma 6 (Terms, Substitutions and Reducts). *With traces$_M$ as denotation function, the traces model has the reduct property stated in Theorem 4.*

As reducts are healthy and the reduct property holds, reducts are well formed. Thanks to Lemma 6 and Theorem 4, the CSP traces model forms an institution.

5   The CSP Stable Failures Model as an Institution

Given an alphabet A the domain $\mathcal{F}(A)$ of the stable failures model consists of those pairs

$$(T, F), \quad \text{where } T \subseteq A^{*\checkmark} \text{ and } F \subseteq A^{*\checkmark} \times \mathcal{P}(A^{\checkmark}),$$

satisfying the following healthiness conditions:

T1 T is non-empty and prefix closed.
T2 $(s, X) \in F \Rightarrow s \in T$.
T3 $s ^\frown \checkmark \in T \Rightarrow (s ^\frown \checkmark, X) \in F$ for all $X \subseteq A^{\checkmark}$.
F2 $(s, X) \in F \wedge Y \subseteq X \Rightarrow (s, Y) \in F$.
F3 $(s, X) \in F \wedge \forall a \in Y : s ^\frown \langle a \rangle \notin T \Rightarrow (s, X \cup Y) \in F$.
F4 $s ^\frown \langle \checkmark \rangle \in T \Rightarrow (s, A) \in F$.

The domain $\mathcal{F}(A)$ gives rise to the notion of stable failures refinement

$$(T, F) \sqsubseteq_{\mathcal{F}} (T', F') :\Leftrightarrow T' \subseteq T \wedge F' \subseteq F$$

$(\mathcal{F}(A), \sqsubseteq_{\mathcal{F}})$ forms a complete lattice with $(A^{*\checkmark}, A^{*\checkmark} \times \mathcal{P}(A^{\checkmark}))$ as its bottom and $(\{\langle\rangle\}, \emptyset)$ as its top. See [18] for a complete definition of the stable failures model. *Morphisms* in the category $Mod_{\mathcal{F}}(A, N)$ are defined as:

$M_1 \rightarrow M_2 :\Leftrightarrow$
$\forall n \in N : \forall a_1, \ldots, a_k \in param(n) : M_2(n(a_1, \ldots, a_k)) \sqsubseteq_{\mathcal{F}} M_1(n(a_1, \ldots, a_k))$

$I(n(a_1, \ldots, a_k)) = (\{\langle\rangle\}, \emptyset)$, i.e. the model which maps all instantiated process names to the denotation of the immediately diverging process, is initial in $Mod_{\mathcal{T}}(A, N)$; $F(n(a_1, \ldots, a_k)) = (A^{*\checkmark}, A^{*\checkmark} \times \mathcal{P}(A^{\checkmark}))$ is final in $Mod_{\mathcal{T}}(A, N)$.
   The semantic clauses of the stable failures model are given by a pair of functions: $fd_M(P) = (traces_M(P), failures_M(P))$ – see [18] for the definition.

Following the same extension pattern for $\alpha : A \to A'$ as demonstrated for the traces model, we obtain:

Theorem 7 (Reducts in the stable failures model are well-behaved)

1. Let $(T', F') \in \mathcal{F}(A')$. Then $\hat{\alpha}(T', F') \in \mathcal{F}(A)$.
2. $\forall X \subseteq A : \hat{\alpha}(\mathcal{F}(\alpha(X))) \subseteq \mathcal{F}(X)$.

Lemma 8 (Terms, Substitutions and Reducts). *With fd_M as denotation function, the stable failures model has the reduct property stated in Theorem 4.*

As reducts are healthy and the reduct property holds, reducts are well formed in the stable failures model. Thanks to Lemma 8 and Theorem 4, the CSP stable failures model forms an institution.

6 Pushouts and Amalgamation

The existence of pushouts and amalgamation properties shows that an institution has good modularity properties. The amalgamation property (called 'exactness' in [9]) is a major technical assumption in the study of specification semantics [20] and is important in many respects. To give a few examples: it allows the computation of normal forms for specifications [3, 5], and it is a prerequisite for good behaviour w.r.t. parametrisation [10] and conservative extensions [9, 17]. The proof system for development graphs with hiding [15], which allow a management of change for structured specifications, is sound only for institutions with amalgamation. A Z-like state based language has been developed over an arbitrary institution with amalgamation [2].

The mildest amalgamation property is that for pushouts. It is also called semi-exactness. An institution is said to be *semi-exact*, if for any pushout of signatures

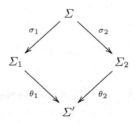

any pair $(M_1, M_2) \in \mathbf{Mod}(\Sigma_1) \times \mathbf{Mod}(\Sigma_2)$ that is *compatible* in the sense that M_1 and M_2 reduce to the same Σ-model can be *amalgamated* to a unique Σ'-model M (i.e., there exists a unique $M \in \mathbf{Mod}(\Sigma')$ that reduces to M_1 and M_2, respectively), and similarly for model morphisms.

Proposition 9. *CspSig does not have pushouts.*

Proof. Suppose that there is a pushout

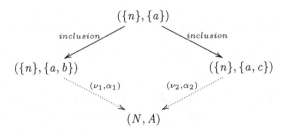

By the pushout property, we have the following mediating morphisms:

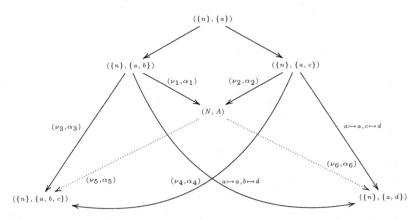

Since α_1 and α_6 are injective, A must have cardinality 2, which implies that α_1 and α_2 are bijective. But then, $\{a, b\} = Im(\alpha_3) = Im(\alpha_5) = Im(\alpha_4) = \{a, c\}$, a contradiction. $\square$

However, this result is not as severe as it might look. Let $CspSig^{noninj}$ be $CspSig$ with the restriction dropped that α must be injective. Then we have:

Proposition 10. $CspSig^{noninj}$ *has pushouts, and any such pushout of a span in* $CspSig$ *actually is a square in* $CspSig$ *(although not a pushout in* $CspSig$).

Proof. **Set** has pushouts, and monomorphisms in **Set** are stable under pushouts ([1, Exercise 11P]). This lifts to the indexed level in $CspSig^{noninj}$ and $CspSig$. $\square$

Note that the phenomenon that pushouts of $CspSig$-spans in $CspSig^{noninj}$ are squares but not pushouts in $CspSig$ is due to the fact that mediating morphisms are generally not in $CspSig$.

Pushouts in $CspSig^{noninj}$ give us an amalgamation property:

Theorem 11. $CspSig^{noninj}$-*pushouts of* $CspSig$-*morphisms have the semi-exactness property for the traces model and the stable failures model.*

Proof. Let

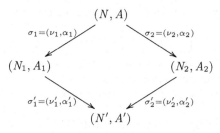

be a *CspSig*^{noninj}-pushout of *CspSig*-morphisms, and let M_i be an (N_i, A_i)-model w.r.t. the trace or the stable failure semantics ($i = 1, 2$) such that $M_1 \mid_{\sigma_1} = M_2 \mid_{\sigma_2}$. We construct an (N', A')-model M' as follows:

$$M'(n) = \begin{cases} \alpha_1(M_1(n_1)), \text{if } n_1 \text{ is such that } \nu_1(n_1) = n \\ \alpha_2(M_2(n_2)), \text{if } n_2 \text{ is such that } \nu_2(n_2) = n \end{cases}$$

This is well-defined because $M_1 \mid_{\sigma_1} = M_2 \mid_{\sigma_2}$. It is clear that $M' \mid_{\theta_i} = M_i$ ($i = 1, 2$). Due to the non-expansion of types principle for signature morphisms, M' is unique. □

In fact, this result generalizes easily to multiple pushouts. Moreover, the initial (=empty) signature has the terminal model category. Since all colimits can be formed by the initial object and multiple pushouts, this shows that we even have exactness (when colimits are taken in *CspSig*^{noninj}).

7   Structuring and Parametrization for CSP

Mostly following [20], in this section we recall a popular set of institution-independent structuring operations, which seems to be quite universal and which can also be seen as a kernel language for the CASL structuring constructs [8].

basic specifications For any signature $\Sigma \in |\mathbf{Sign}|$ and finite set $\Gamma \subseteq \mathbf{Sen}(\Sigma)$ of Σ-sentences, the *basic specification* $\langle \Sigma, \Gamma \rangle$ is a specification with:

$Sig(\langle \Sigma, \Gamma \rangle)\ :=\ \Sigma$
$\mathbf{Mod}(\langle \Sigma, \Gamma \rangle) := \{M \in \mathbf{Mod}(\Sigma) \mid M \models \Gamma\}$

union: For any signature $\Sigma \in |\mathbf{Sign}|$, given Σ-specifications SP_1 and SP_2, their *union* $SP_1 \cup SP_2$ is a specification with:

$Sig(SP_1 \cup SP_2)\ :=\ \Sigma$
$\mathbf{Mod}(SP_1 \cup SP_2) := \mathbf{Mod}(SP_1) \cap \mathbf{Mod}(SP_2)$

translation: For any signature morphism $\sigma \colon \Sigma \to \Sigma'$ and Σ-specification SP, SP **with** σ is a specification with:

$Sig(SP \text{ \textbf{with} } \sigma)\ :=\ \Sigma'$
$\mathbf{Mod}(SP \text{ \textbf{with} } \sigma) := \{M' \in \mathbf{Mod}(\Sigma') \mid M' \mid_{\sigma} \in \mathbf{Mod}(SP)\}$

hiding: For any set SYs of symbols in a signature Σ' generating a subsignature Σ of Σ', and Σ'-specification SP', SP' **reveal** SYs is a specification with:

$\quad\quad Sig(SP'$ **reveal** $SYs)\quad := \Sigma$

$\quad\quad\mathbf{Mod}(SP'$ **reveal** $SYs) := \{M'\mid_\sigma \mid M' \in \mathbf{Mod}(SP')\}$

where $\sigma\colon \Sigma \to \Sigma'$ is the inclusion signature morphism.

free specification: For any signature morphism $\sigma\colon \Sigma \to \Sigma'$ and Σ'-specification SP', **free** SP' **along** σ is a specification with:

$Sig(\mathbf{free}\ SP'\ \mathbf{along}\ \sigma) = \Sigma'$

$\mathbf{Mod}(\mathbf{free}\ SP'\ \mathbf{along}\ \sigma) = \{M' \in \mathbf{Mod}(SP') \mid$

$\quad\quad M'$ is strongly persistently $(\mathbf{Mod}(\sigma)\colon \mathbf{Mod}(SP') \longrightarrow \mathbf{Mod}(\Sigma))$-free $\}$

Given categories $\mathbf{A}$ and $\mathbf{B}$ and a functor $G\colon \mathbf{B} \longrightarrow \mathbf{A}$, an object $B \in \mathbf{B}$ is called _G-free (with unit $\eta_A\colon A \longrightarrow G(B)$) over_ $A \in \mathbf{A}$, if for any object $B' \in \mathbf{B}$ and any morphism $h\colon A \longrightarrow G(B')$, there is a unique morphism $h^\#\colon B \longrightarrow B'$ such that $G(h^\#) \circ \eta_A = h$. An object $B \in \mathbf{B}$ is called _strongly persistently G-free_ if it is G-free with unit id over $G(B)$ (_id_ denotes the identity).

parametrisation: For any (formal parameter) specification SP, (body) specification SP' with signature inclusion $\sigma\colon Sig(SP) \longrightarrow Sig(SP')$ and specification name SN, the declaration

$$SN[SP] = SP'$$

names the specification SP' with the name SN, using formal parameter SP. The formal parameter SP can also be omitted; in this case, we just have named a specification for future reference.

instantiation: Given a named specification $SN[SP] = SP'$ with signature inclusion $\sigma\colon Sig(SP) \longrightarrow Sig(SP')$ and an (actual parameter) specification SPA and a fitting morphism $\theta\colon Sig(SP) \longrightarrow Sig(SPA)$, $SN[SPA$ **fit** $\theta]$ is a specification with

$\quad\quad Sig(SN[SPA\ \mathbf{fit}\ \theta])\quad := \Sigma$

$\quad\quad\mathbf{Mod}(SN[SPA\ \mathbf{fit}\ \theta]) :=$

$\quad\quad\quad \{M \in \mathbf{Mod}(\Sigma) \mid M\mid_{\sigma'} \in \mathbf{Mod}(SP'), M\mid_{\theta'} \in \mathbf{Mod}(SPA)\}$

where

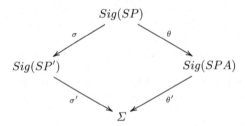

is a pushout (note that for CSP, we take the pushout in $CspSig^{noninj}$, as discussed in Sect. 6).

In CASL, we can also _extend_ specifications with new declarations and axioms. This is written SP **then** SP', where SP' is a specification fragment. Since we do not want to deal with specification fragments formally here, we just note that

spec NoLoss =
- $OneCoin = coin \rightarrow Skip$
- $AtLeastOneCoin = OneCoin \sqcap coin \rightarrow AtLeastOneCoin$
- $NoLoss = AtLeastOneCoin;\ item \rightarrow NoLoss$

end

spec MachineForTeaAndCoffee
  [{NoLoss **then** $NoLoss \sqsubseteq VM \setminus \{button\}$ } **reveal** VM]
=
  $TeaAndCoffee =\quad VM[[\{(item, coffee), (button, c\text{-}button)\}]]$
  $\qquad\qquad\qquad \Box\ VM[[\{(item, tea), (button, t\text{-}button)\}]]$

end

spec UnfairMachine =
  $UnfairMachine = button \rightarrow coin \rightarrow coin \rightarrow item \rightarrow UnfairMachine$
end

spec UnfairMachineForTeaAndCoffee =
  { MachineForTeaAndCoffee [UnfairMachine **fit** $UnfairMachine \mapsto VM$]
  } **reveal** TeaAndCoffee
end

Fig. 6. Process Instantiation

the semantics of extension is similar to that of union, and refer to [8] for full formal details.

In standard CSP, the cpo approach defines the meaning of a system of recursive process equations to be its smallest fixed-point, if such a smallest fixed-point exists. To determine this fixed point, Tarski's fixed-point theorem is applied to the function underlying the system of equations. Take for example, the system $P = P$, $Q = a \rightarrow Q$. Over the alphabet $A = \{a\}$ it has $traces(P) = \{\langle\rangle\}, traces(Q) = a^*$ as its smallest solution. However, there are other fixed-points, as the equation $P = P$ holds for every process, i.e. $traces(P) = \{\langle\rangle, \langle a\rangle\}$, $traces(P) = \{\langle\rangle, \langle a\rangle, \langle aa\rangle\}$, etc. also yield fixed-points. As structured CSP works with loose semantics,

spec Loose = • $P = P$ • $Q = a \rightarrow Q$ **end**

has the set of *all* fixed-points as its semantics. Choosing initial semantics by adding the keyword **free**, however, i.e.

spec Initial = **free** { • $P = P$ • $Q = a \rightarrow Q$ } **end**

has the *smallest* fixed point as its semantics thanks to our choice of morphisms in the model categories.

In order to illustrate the practical use of structured CSP specifications, we consider the classical example of process algebra: the development of a vending machine for tea and coffee, following [13], see Fig. 6. For simplicity, we omit explicit signature declarations and derive the alphabet and the process names from the symbols used. The owner of a vending machine will insist the machine never to make a loss. The process *NoLoss* with $sort(NoLoss) = \{coin, item\}$ in the specification NoLoss has the property that at any time the number of *coins* inserted to the machine is bigger than the number of *items* delivered. The specification MACHINEFORTEAANDCOFFEE describes how to turn the specification of a non-dedicated vending machine *VM* into the specification of a machine for selling tea and coffee. Here, we assume $sort(VM) = \{coin, item, button\}$. *VM* is loosely specified by the condition $NoLoss \sqsubseteq VM \setminus \{button\}$, i.e. $VM \setminus \{button\}$ does not make any loss. The specification MACHINEFORTEAANDCOFFEE takes the machine *VM* as its parameter and defines the machine *TeaAndCoffee* by renaming the *item* to be delivered into *tea* and *coffee*, resp., and the *button* into *c-button* and *t-button*, resp. However, only those vending machines *VM* are accepted as an actual parameter that fulfil the condition specified by *NoLoss*: This is expressed via the refinement condition $NoLoss \sqsubseteq VM \setminus \{button\}$ in the parameter[7]. The *UnfairMachine*, which lets the customer pay twice for one item, fulfils this requirement in the traces model as well as in the stable failures model. Therefore, it is a legal parameter. Instantiating MACHINEFORTEAAND-COFFEE with the process *UnfairMachine* yields a process *CoffeeAndTea*, where the customer has to pay twice for tea and coffee.

The semantics of the specifications above behaves as expected. For example, for the basic specification NoLoss, we get:

- $Sig(\text{NoLoss}) = (A, (\bar{N}, sort, param))$ with
- $A = \{coin, item\}$
- $\bar{N} = \{OneCoin, AtLeastOneCoin, NoLoss\}$
- $sort(OneCoin) = \{coin\}$, $sort(AtLeastOneCoin) = \{coin\}$, $sort(NoLoss) = A$.
- $param(n) = \langle\rangle$.
- **Mod**(NoLoss) consists of one model M with

$$M(NoLoss) = \{s \mid s \text{ is prefix of } t \in (coin^+ \ item)^*\}$$

It is quite typical that CSP specifications have exactly one model; indeed, in this respect, CSP resembles more a programming language than a specification language. However, using refinement, we can also write useful loose specifications. Consider $SP = \{\text{NoLoss } \textbf{then } NoLoss \sqsubseteq VM \setminus \{button\}\} \textbf{ reveal } VM$. This has the following semantics:

- $Sig(SP) = (A, (\bar{N}, sort, param))$ with
- $A = \{coin, item, button\}$
- $\bar{N} = \{VM\}$
- $sort(VM) = \{coin, item, button\}$.
- $param(n) = \langle\rangle$.
- **Mod**(SP) consists of models M that provide a trace set $M(VM)$ with the following property: if the action *button* is removed from $M(VM)$, the resulting trace set is contained in $M(NoLoss)$ above.

---

[7] Note that CSP refinement $P \sqsubseteq Q$ is equivalent to $P = P \sqcap Q$.

That is, SP can be see as a requirement specification on a vending machine, allowing several actual vending machine implementations. SP is the formal parameter of a parametrised specification that can be instantiated with different vending machines. Moreover, due to the amalgamation property of Theorem 11, we can ensure that each vending machine model can be extended to a model of the appropriately instantiated specification MACHINEFORTEAANDCOFFEE.

8   Conclusion and Future Work

Our institutions for CSP use injective signature morphisms, due to the fact that the alphabet plays a double role, in the process syntax and the semantic domains, and both aspects are mapped covariantly — a contravariant mapping would destroy important laws for of CSP processes.

Languages like Unity and CommUnity [11] split the alphabet of communications into 'data' – to be translated covariantly – and 'actions' – to be translated contravariantly. The advantage of this approach is that the contravariant translation makes it possible to 'split' actions. We avoid such a partition of the alphabets of communications as CSP with its relational renaming already offers a means of 'splitting' an action on the term level. A rich set of algebraic laws allows to relates the new process with the old one.

We have demonstrated that with our CSP institutions, structured specifications have a semantics that fits with what one would expect in the CSP world. In particular, we can use loose semantics and parameterisation in combination with CSP refinement in a very useful way, going beyond what has been developed in the CSP community so far. Future work will extend the institutions presented here with an algebraic data type part, aiming at an institution for the language CSP-CASL [16]. For this, it is probably useful to distinguish between a syntactic and a semantic alphabet, at the price of complicating algebraic laws like the $\langle \Box\text{-step}\rangle$ law by using equality on the semantic alphabet in a subtle way, but with the advantage of allowing for non-injective alphabet translations.

Acknowledgment. The authors would like to thank José Fiadeiro, Yoshinao Isobe, Grigore Rosu, Erwin R. Catesbeiana, Andrzej Tarlecki, and Lutz Schröder for helpful discussions on what the right CSP institution might be.

References

1. J. Adámek, H. Herrlich, and G. Strecker. *Abstract and Concrete Categories*. Wiley, New York, 1990.
2. H. Baumeister. *Relations between Abstract Datatypes modeled as Abstract Datatypes*. PhD thesis, Universität des Saarlandes, 1998.
3. J. Bergstra, J. Heering, and P. Klint. Module algebra. *Journal of the ACM*, 37:335–372, 1990.
4. M. Bidoit and P. D. Mosses. CASL *User Manual*. LNCS 2900. Springer, 2004.
5. T. Borzyszkowski. Logical systems for structured specifications. *Theoretical Computer Science*, 286:197–245, 2002.

6. B. Buth, J. Peleska, and H. Shi. Combining methods for the livelock analysis of a fault-tolerant system. In *AMAST'98*, LNCS 1548. Springer, 1998.

7. B. Buth and M. Schrönen. Model-checking the architectural design of a fail-safe communication system for railway interlocking systems. In *FM'99*, LNCS 1709. Springer, 1999.

8. CoFI (The Common Framework Initiative). CASL *Reference Manual*. LNCS 2960. Springer Verlag, 2004.

9. R. Diaconescu, J. Goguen, and P. Stefaneas. Logical support for modularisation. In *Logical Environments*, pages 83–130. Cambridge, 1993.

10. H. Ehrig and B. Mahr. *Fundamentals of Algebraic Specification 2*. Springer, 1990.

11. J. Fiadeiro. *Categories for Software Engineering*. Springer, 2004.

12. J. Goguen and R. Burstall. Institutions: Abstract model theory for specification and programming. *Journal of the ACM*, 39:95–146, 1992.

13. C. A. R. Hoare. *Communicating Sequential Processes*. Prentice Hall, 1985.

14. Y. Isobe and M. Roggenbach. A complete axiomatic semantics for the CSP stable-failures model. In *CONCUR'06*, LNCS 4137. Springer, 2006.

15. T. Mossakowski, S. Autexier, and D. Hutter. Development graphs – proof management for structured specifications. *Journal of Logic and Algebraic Programming*, 67(1-2):114–145, 2006.

16. M. Roggenbach. CSP-CASL - a new integration of process algebra and algebraic specification. *Theoretical Computer Science*, 354(1):42–71, 2006.

17. M. Roggenbach and L. Schröder. Towards trustworthy specifications I: Consistency checks. In *WADT'01*, LNCS 2267. Springer, 2001.

18. A. W. Roscoe. *The Theory and Practice of Concurrency*. Prentice Hall, 1998.

19. P. Ryan, S. Schneider, M. Goldsmith, G. Lowe, and B. Roscoe. *The Modelling and Analysis of Security Protocols: the CSP Approach*. Addison-Wesley, 2001.

20. D. Sannella and A. Tarlecki. Specifications in an arbitrary institution. *Information and Computation*, 76:165–210, 1988.

Incremental Resolution of Model Inconsistencies

Tom Mens[1,2] and Ragnhild Van Der Straeten[3]

[1] Software Engineering Lab, Université de Mons-Hainaut
Av. du champ de Mars 6, 7000 Mons, Belgium
tom.mens@umh.ac.be
[2] LIFL (UMR 8022), Université Lille 1 - Projet INRIA Jacquard
Cité Scientifique, 59655 Villeneuve d'Ascq Cedex, France
[3] Systems and Software Engineering Lab, Vrije Universiteit Brussel
Pleinlaan 2, 1050 Brussel, Belgium
rvdstrae@vub.ac.be

Abstract. During model-driven software development, we are inevitably confronted with design models that contain a wide variety of inconsistencies. Interactive and automated support for detecting and resolving these inconsistencies is therefore indispensable. In this paper, we report on an iterative inconsistency resolution process. Our approach relies on the underlying formalism of graph transformation. We exploit the mechanism of critical pair analysis to analyse dependencies and conflicts between inconsistencies and resolutions, to detect resolution cycles and to analyse the completeness of resolutions. The results of this analysis are integrated in the iterative inconsistency resolution process and can help the software engineer to develop and evolve models in presence of inconsistencies.

1 Introduction

During development and evolution of analysis and design models it is often desirable to tolerate inconsistencies in design models [1]. Such inconsistencies are inevitable for many reasons: (i) in a distributed and collaborative development setting, different models may be developed in parallel by different persons; (ii) the interdependencies between models may be poorly understood; (iii) the requirements may be unclear or ambiguous; (iv) the models may be incomplete and subject to interpretation because some essential information is intentionally left out to avoid premature decisions; (v) the models are continuously subject to evolution; (vi) the semantics of the modeling language itself may be poorly specified.

All of these reasons can hold in the case of the Unified Modeling Language (UML), the de-facto general-purpose modelling language [2]. Therefore, current UML modeling tools should provide better support for dealing with inconsistencies. These inconsistencies may either be localised in a single UML diagram, or spread over different (types of) UML diagrams. An example of the latter is illustrated in Fig. 1. The behaviour of the *CardReader* class (belonging to some class diagram) is specified by a protocol state machine that refers to an operation *retainCard()* that is not defined in the class *CardReader* or any of its ancestors. We call this situation a *Dangling Operation Reference*.

J.L. Fiadeiro and P.-Y. Schobbens (Eds.): WADT 2006, LNCS 4409, pp. 111–126, 2007.
© Springer-Verlag Berlin Heidelberg 2007

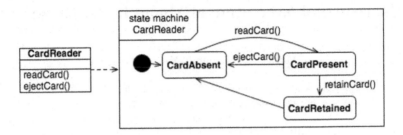

Fig. 1. Example of a *Dangling Operation Reference* inconsistency between a class diagram and protocol state machine

In this paper, we will introduce a Simple Iterative Inconsistency Resolution Process (SIRP). This process provides a formally founded, yet automated, approach to detect and resolve model inconsistencies. Our approach relies on graph transformation theory. It detects inconsistencies automatically and proposes possible resolutions to the user.

Our approach contributes to the management of the different detected inconsistencies and their possible resolutions. A first step to manage the possibly large amount of detected inconsistencies and resolutions is to analyse the parallel conflicts and sequential dependencies between inconsistencies and resolutions. A second step is to ascertain that all possible resolutions for a given inconsistency are effectively treated. In this article we show and discuss how graph transformation dependency analysis can be exploited to achieve these goals and we report on a tool we are developing to support the inconsistency resolution process.

2 Simple Iterative Resolution Process

We implemented tool support for an iterative and incremental process of resolving model inconsistencies. First, inconsistencies in the model are identified. Next, resolutions are proposed, selected and applied. The user may also wish to ignore or disable certain types of inconsistencies or resolutions. This process continues until all problems are resolved or until the user is satisfied.

A screenshot of the tool that we developed to support this process is shown in Fig. 2.[1] Several inconsistencies have been resolved already, as shown in the resolution history. For an occurrence of the *dangling operation reference* inconsistency, four resolutions are proposed with a certain popularity (based on whether the rule has been applied before by the user). Selected resolutions can be applied to resolve the selected inconsistency.

Because model inconsistency resolution is an inherently incremental and iterative process, a number of typical problems may arise:

- **Induced inconsistencies.** The resolution of a certain inconsistency may induce other inconsistencies as a side effect.
- **Conflicting resolutions.** Some resolutions may not be jointly applicable, even if they resolve another type of inconsistency.

---

[1] The tool was developed by Jean-François Warny during his Masters Thesis [3].

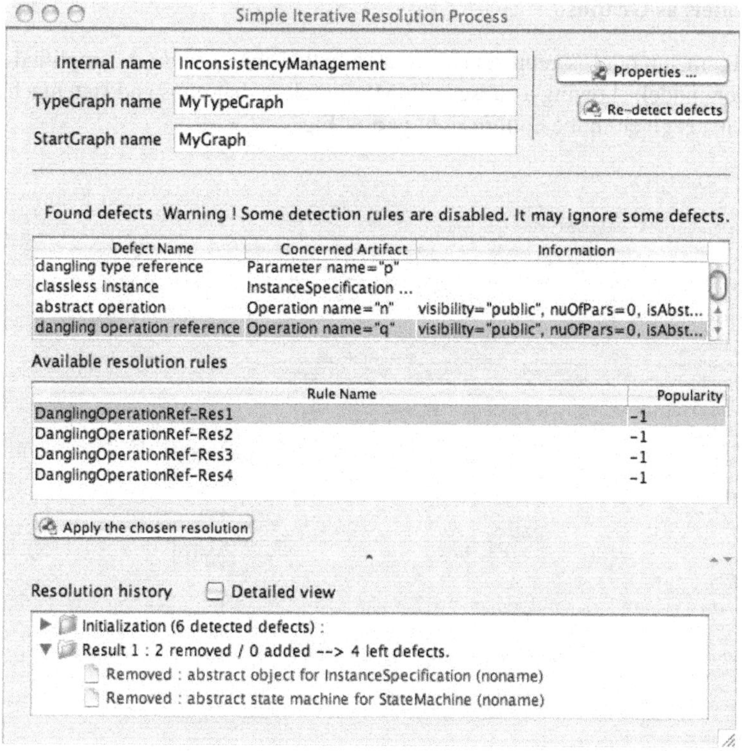

Fig. 2. Screenshot of the incremental inconsistency resolution tool in action

- **Resolution cycles.** Certain sequences of resolutions may reintroduce an inconsistency that has been resolved earlier.

During the inconsistency resolution process, it is very important to be aware of such situations, and to take the appropriate actions when they arise. Because, in practice, the set of model inconsistencies and their resolutions can be quite large, it is virtually impossible to do this analysis manually. For this reason, automated support is needed. The next section explains the formal approach we adopted and support by our tool.

3 Inconsistencies and Resolutions as Graph Transformation Rules

To specify model inconsistencies and their resolutions, we opted for a formal description based on the theory of graph transformation [4,5]. The main benefit is that this allows us to rely on theoretical results about critical pairs [6] to perform analysis of parallel conflicts and sequential dependencies between graph transformation rules. In this article, we exploit this feature to optimise the resolution process, by analysing and detecting conflicts and dependencies between resolutions. We also use the technique to check to which extent the provided set of resolutions is complete.

3.1   Models as Graphs

The UML metamodel is represented by a so-called *type graph*. A simplified version of the metamodel, showing a subset of UML 2.0 class diagrams and statemachine diagrams only, is given in the bottom right part of Fig. 3.

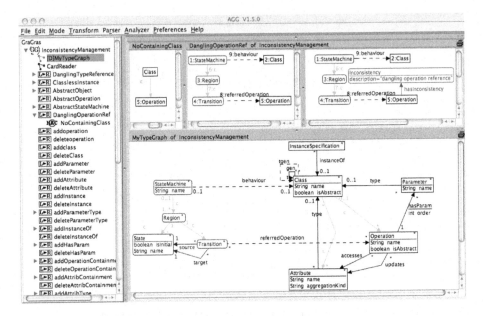

Fig. 3. Screenshot of the AGG tool. The bottom right graph displays the type graph, corresponding to a simplified version of the UML metamodel.

A UML model will be represented as a *graph* that satisfies the constraints imposed by the aforementioned *type graph*. Fig. 4 shows a directed, typed, attributed graph that represents the UML model of Fig. 1. This graph representation can be generated automatically from the corresponding UML model without any loss of information. [2]

3.2   Detecting Model Inconsistencies with Graph Transformation Rules

Model inconsistencies can be detected automatically by means of *graph transformation rules*. For each type of inconsistency, a graph transformation rule is specified that detects the inconsistency. This is realised by searching for the occurrence of certain graph structures in the model, as well as the *absence* of certain forbidden structures (so-called *negative application conditions* or NACs). When an inconsistency is found, the graph is annotated with a new node of type *Inconsistency* pointing to the node that is the source of this inconsistency. An example is given in Fig. 4, where an occurrence of the

---

[2] An experiment along these lines has been carried out by Laurent Scolas as a student project.

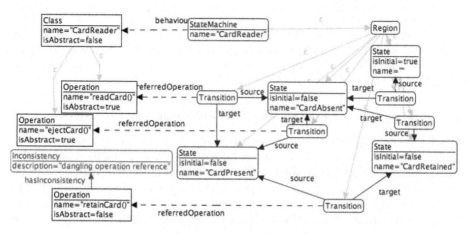

Fig. 4. Graph representation of the inconsistent UML model of Fig. 1

Dangling Operation Reference inconsistency has been found, as can be seen in the *description* attribute of the *Inconsistency* node. In the remainder of this article, detection rules denote the graph transformation rules expressing an inconsistency.

The top right part of Fig. 3 gives the formal specification of the detection rule for the *Dangling Operation Reference* inconsistency. This occurs when a message specified on a transition in a protocol state machine does not correspond to an existing *Operation* defined in some existing *Class*. The specification of this rule as a graph transformation is composed of three parts. The middle pane represents the *left-hand side (LHS)* of the rule. The leftmost pane represents a *negative application condition (NAC)*, expressing the fact that the *Operation* of interest is not defined in any existing *Class*. Finally, the rightmost pane represents the *right-hand side (RHS)* of the rule, showing the result after the transformation. In this case, the only modification is the introduction of a *Inconsistency* node that is linked to the *Operation* to indicate a model inconsistency has been detected.

Given a source model, we can apply all detection rules in sequence to detect all possible model inconsistencies. Different occurrences of the same inconsistency may be detected at different locations, and the same model element may be annotated with occurrences of multiple inconsistencies. By construction, the detection rules are parallel independent, i.e., the application of a detection rule has no unexpected side effects on other detection rules. This is because a detection rule only introduces in its RHS a new node of type *Inconsistency*. Moreover, the LHS and NAC of a detection rule never contain any *Inconsistency* nodes.

As a technical side note, we needed to avoid repeated application of an inconsistency rule for the same match, since this could potentially introduce arbitrarily many copies of the *Inconsistency* node. This problem was tackled as follows. During the process of finding a match of the LHS of the rule in the source model, matches that have been used in previous applications of the rule are excluded automatically. This process works because the inconsistency detection rules are monotonous and only add new information.

An alternative way to achieve the same thing would be to include the *Inconsistency* node explicitly as a NAC in each of the inconsistency detection rules. We decided not to do it in this way because it would make the rules more complex and because it would cause undesired side effects during the resolution analysis process that will be explained in Sect. 4.

3.3 Resolving Model Inconsistencies with Graph Transformation Rules

Graph transformation rules are also used to *resolve* previously detected inconsistencies. In the remainder of this article, resolution rules denote the graph transformation rules expressing an inconsistency resolution. For each type of model inconsistency, several resolution rules can be specified. Each resolution rule has the same general form. On the LHS we find an *Inconsistency* node that indicates the particular inconsistency that needs to be resolved. On the RHS this *Inconsistency* node has disappeared because the rule removes the inconsistency.

As an example, Fig. 5 proposes three resolution rules for the *Dangling Operation Reference* inconsistency. Note that other resolution rules are also possible.

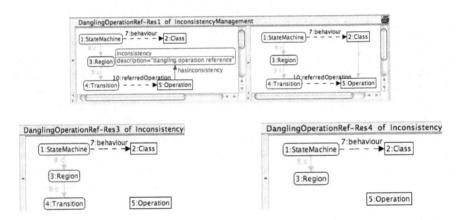

Fig. 5. Three graph transformations specifying alternative resolution rules for the *Dangling Operation Reference* inconsistency. For the last two resolution rules, only the RHS is shown, since the LHS is the same as for the first resolution rule.

3.4 AGG

The tool that we have used to perform our experiments is AGG^3 (version 1.5.0), a state-of-the-art general purpose graph transformation tool [7]. It provides direct support for all concepts mentioned before: directed, attributed, typed graphs and typed graph transformations. In addition to the *type graph*, extra *graph constraints* can be expressed in

---

[3] See http://tfs.cs.tu-berlin.de/agg/. Note that the screenshot shown in Fig. 2 is not part of the standard AGG distribution. It is an extension of the AGG tool that we have developed for the purposes of this article.

AGG, and the consistency of a graph with respect to these graph constraints can be checked. *Termination* of certain types of graph grammars can be checked automatically by *AGG*.

Critical pair analysis is provided by *AGG* to detect critical pairs between graph transformation rules [6,8]. The goal is to compute all potential mutual exclusions and sequential dependencies for a given set of transformation rules by pairwise comparison. Critical pair analysis of graph transformations has already been used to detect conflicting functional requirements in UML models composed of use case diagrams, activity diagrams and collaboration diagrams in other, related, domains in [9]. In previous work [10], we used it to detect conflicts and dependencies between software refactorings. Recently we explored its use in the context of model inconsistency management [11]. The current article is a continuation of this line of research. In the future, we will also make use of the other formal reasoning mechanisms provided by *AGG*.

4 Inconsistency Resolution Analysis

As already mentioned in Sect. 2, model inconsistency resolution is an incremental and iterative process in which induced inconsistencies, conflicting resolutions and resolution cycles can occur. In this section, we explain how we exploited the technique of *critical pair analysis* to achieve automated support for the aforementioned problems. Such analysis is directly supported by the *AGG* engine, so it can readily be used in SIRP[4]. For a detailed explanation of the inconsistencies and resolution used in the remainder of this paper we refer to [11] in which we presented a set of inconsistencies and resolutions.

4.1 Induced Inconsistencies

Induced inconsistencies may appear when the resolution of a certain model inconsistency introduces one or more other inconsistencies as a side effect. For example, suppose that we have a model that contains an inconsistency of type *dangling operation reference*, and we want to resolve this problem using the first resolution rule of Fig. 5, i.e., by adding a containment relationship between an existing class and an existing operation. If, however, the class was concrete, while the operation was abstract, the inconsistency resolution induces a new type of model inconsistency called *abstract operation*. This is because a concrete class is not supposed to contain any abstract operations.

The problem of *induced inconsistencies* is a typical situation of a *sequential dependency*: an inconsistency detection rule causally depends on a previously applied resolution rule. Fig. 6 shows an example of a dependency graph that has been generated by AGG. It shows all resolution rules that may induce a *Dangling Operation Reference* inconsistency. For example, we see that this inconsistency sequentially depends on *AbstractOperation-Res4*, the specification of which is given in the top right part of Fig. 7. This kind of information is quite important in an incremental resolution process,

---

[4] The current version of SIRP does not yet incorporate these results, integration of critical pair analysis in SIRP is scheduled for a later version of the tool.

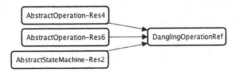

Fig. 6. Dependency graph generated by AGG showing how the inconsistency of type *Dangling Operation Reference* may be induced by different resolution rules

as it informs us which types of inconsistencies will need to be redetected after the application of a given resolution rule.

4.2 Conflicting Resolutions

Conflicting resolutions may appear when there are multiple inconsistencies in a model, each having their own set of applicable resolution rules. It may be the case that applying a resolution rule for one inconsistency, may invalidate another resolution rule for another inconsistency. As an example, consider Fig. 7. The left pane depicts a situation where two inconsistencies occur, one of type *Abstract Operation* and *Dangling Operation Reference* respectively, but attached to different model elements. The resolution rules *AbstractOperation-Res4* and *DanglingOperationRef-Res2* for these inconsistencies (shown on the right of Fig. 7) are conflicting, since the first resolution rule sets the relation *contains* connecting class 1 to operation 2 to connecting class 4 and operation 2, whereas the second resolution rule requires as a precondition that class 1 is connected to operation 2 through a containment relation.

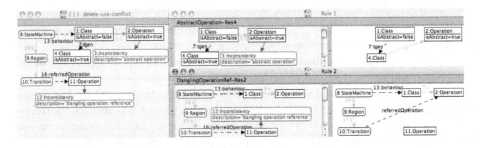

Fig. 7. *Conflicting resolutions*: Example of a critical pair illustrating a mutual exclusion between resolution rules *AbstractOperation-Res4* and *DanglingOperationRef-Res2*

The problem of *conflicting resolutions* is a typical situation of a *parallel conflict*: two rules that can be applied independently cannot be applied one after the other (i.e., they are mutually exclusive) because application of the first rule prevents subsequent application of the second one. This kind of information is quite important during an interactive resolution process, as it informs the user about which resolution rules are mutually exclusive and, hence, cannot be applied together.

4.3   Resolution Cycles

Starting from the dependency graph, we can also compute possible *cycles* in the conflict resolution process. This may give important information to the user (or to an automated tool) to avoid repeatedly applying a certain combination of resolution rules over and over again. Clearly, such cycles should be avoided, in order to optimise the resolution process. AGG allows us to compute a conservative approximation of such cycles. This is illustrated in Fig. 8, which represents a carefully selected subset of sequential dependencies that have been computed by AGG.[5] In this figure, we observe the presence of multiple cycles of various lengths, all of them involving the *Abstract Operation* inconsistency.

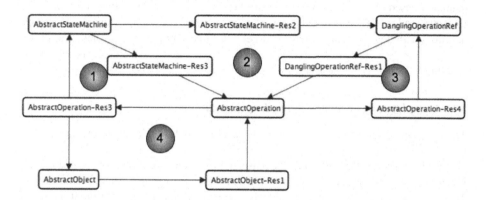

Fig. 8. Some examples of detected cycles in the sequential dependency graph

Let us start by analysing the cycles of length 4, that correspond to an alternation of two successive detection and resolution steps. The cycle corresponding to region 1 shows that we can repeatedly apply resolution rules *AbstractStateMachine-Res3* and *AbstractOperation-Res3 ad infinitum*. This is the case because the two resolution rules are inverses of each other. Therefore, after applying one of both rules, a resolution tool should not propose the other rule in the same context, as it would undo the effect of the first one. The cycle corresponding to region 4 is similar to the previous one, except that it occurs between resolution rules *AbstractObject-Res1* and *AbstractOperation-Res3*.

There is also a cycle of length 6, corresponding to a succession of three detection and resolution rules. The cycle is described by the boundaries of the region composed by 1 and 2, and occurs when we apply resolution rules *AbstractStateMachine-Res2*, *DanglingOperationRef-Res1* and *AbstractOperation-Res3* in sequence.

One should note that the computed dependency graph only presents a conservative approximation of what can actually happen in a concrete setting. Sometimes, false positives may be reported by the dependency analysis. Careful manual analysis is required to reveal such situations. An example is the cycle corresponding to region 3. It turns out

---

[5] To interpret the dependency graph, the blue directed arrows should be interpreted as "enables" or "triggers".

that the sequential dependency between *AbstractOperation-Res4* and *DanglingOperationRef* is a false positive. A more sophisticated dependency analysis algorithm would be needed to resolve this problem.

Because the sequential dependency graph can be very large, manual detection of cycles is unfeasible in practice. Therefore, we have used a small yet intuitive user interface for detecting all possible cycles in a flexible and interactive way, based on the output generated by AGG's critical pair analysis algorithm. [6]

5 Completeness

The resolution rules and inconsistencies proposed in this article have been specified manually based on our intuition and on similar work carried out by other authors. These resolution rules correspond to typical ways to resolve the inconsistencies. However, nothing guarantees us that we didn't forget any important resolution rules. Therefore, this section discusses *completeness* issues related to our approach to inconsistency management.

5.1 Completeness of Inconsistency Representation

There is a first question related to completeness for which we already know the answer: *Can we represent all possible types of model inconsistencies as graph transformation rules?* The answer is "no" for multiple reasons. First, there are certain types of semantic inconsistencies that are quite hard to express by means of graph transformation rules. Therefore, we focus on structural inconsistencies only. Second, the notion of inconsistency is ill-defined. There is a virtually infinite set of things that can be regarded as an inconsistency, because it is a subjective notion. Different persons tend to have a different interpretation of what it means for something to be inconsistent. Of course, in theory, it is possible to provide a strict and precise definition of (in)consistency. In *AGG*, syntactically inconsistent models are defined by means of their type graph: every graph that does not correspond to the constraints imposed by the type graph is considered to be inconsistent. For our purposes, however, this approach is insufficient. There are many types of inconsistencies that we want to detect but that cannot be expressed by relying on the type graph alone. Additional graph constraints are needed to cover these cases.

5.2 Completeness of Resolution Rules

Resolution rules consist of –what we call– *primitive* operations performed on the user-defined model. Primitive operations are the addition or removal of a given type of model element, or changing the attribute values of a given model element (e.g., its name) or its references (e.g., the containment relation between an operation and a class).

For the type graph of Fig. 3, we have expressed each of these primitive operations manually as (parameterised) graph transformation rules in AGG. In the future, we will automate the process of generating the transformation rules corresponding to primitive

---

[6] This program has been developed by Stéphane Goffinet in the course of a student project. Currently, the tool is not yet integrated in SIRP, this is planned for a next version.

operations. Some of these rules are shown on the left of Fig. 3: *addOperation, delete-Operation, addClass, deleteClass* and so on.

With respect to the inconsistency resolution rules, we have asked ourselves the following two relevant questions: (1) Given a particular model inconsistency, what is the complete set of primitive operations that are needed to resolve the inconsistency? (2) Given a particular primitive operation, what are all the model inconsistencies that it may resolve?

These questions pertain to the relation between primitive operations and inconsistency resolution rules. They can be answered by critical pair analysis. The reasoning is as follows. By computing all parallel conflicts between inconsistency detection rules and primitive operations, we find all situations in which a primitive operation invalidates at least one of the (positive or negative) preconditions of the inconsistency. As a result, after applying the primitive operation, the inconsistency will no longer be present. This implies that any inconsistency resolution rule for this particular inconsistency must include at least one of these primitive operations.

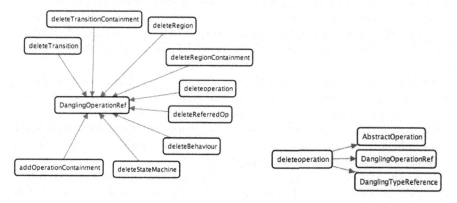

Fig. 9. Parallel conflicts between primitive operations and inconsistencies. On the left, we see all possible primitive operations that may potentially resolve the *Dangling Operation Reference* inconsistency. On the right, we see all possible inconsistencies that may be resolved by the *deleteOperation* primitive.

In response to question (1), consider the left part of Fig. 9 as an example. It shows all parallel conflicts between the *Dangling Operation Reference* inconsistency and the primitive operations. We conclude from this figure that the inconsistency may be (at least partially) resolved by applying a *deleteTransition* operation, a *deleteReferredOp* operation, an *addOperationContainment* operation, and so on. Coming back to the resolution rules that we manually specified in Fig. 5 based on our intuition, we see that this reasoning is correct, since each of these resolution rules involves at least one of the primitive operations identified in Fig. 9. Rule *DanglingOperationRef-Res1* corresponds to the application of an *addOperationContainment* operation. Rule *Dangling-OperationRef-Res3* corresponds to the application of a *deleteReferredOp* operation.

Rule *DanglingOperationRef-Res4* involves a combination of several of the primitive operations that we identified: *deleteTransition, deleteTransitionContainment,* and *deleteReferredOp.*

In a similar vein, every resolution rule for the *Dangling Operation Reference* inconsistency will involve at least one of the primitive operations identified in Fig. 9. As such, we can use this analysis to assess whether a set of resolution rules that has been specified for a given model inconsistency is complete, in the sense that all possible situations have been considered. Such an analysis is quite useful, since it allows us to find the primitive operations that need to be included in resolution rules of a certain inconsistency.

To answer question (2), consider the right part of Fig. 9 as an example. It shows inconsistencies that the *deleteOperation* primitive may potentially resolve. The same kind of information can be generated automatically for every kind of primitive operation.

5.3   Completeness of Detection Rules

With respect to the inconsistency detection rules, we have asked ourselves two similar questions. (1) Given a particular model inconsistency, can we identify the complete set of primitive operations that potentially give rise to this inconsistency? (2) Given a particular primitive operation, what are all the model inconsistencies that it potentially gives rise to?

As described in Sect. 5.2, the primitive operations can be derived from the type graph and can be expressed as (parameterised) graph transformation rules. The questions stated above pertain to the relation between primitive operations and inconsistency detection rules. They can be answered by computing all sequential dependencies between both types of rules. In fact, they give us an alternative view on the *induced inconsistencies* that have been discussed in Sect. 4.1.

In response to question (1), consider the left part of Fig. 10 as an example. It shows all sequential dependencies between the *Dangling Operation Reference* inconsistency

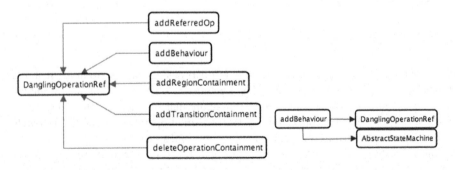

Fig. 10. Sequential dependencies between primitive operations and inconsistencies. On the left, we see all possible primitive operations that may potentially induce the *Dangling Operation Reference* inconsistency. On the right, we see all possible inconsistencies that may be induced by the *addBehaviour* primitive operation.

and the primitive operations. We see that the inconsistency may be induced by different types of primitive operations. Whenever one of the primitive operations is applied, a new occurrence of the inconsistency may be introduced. For example, the inconsistency may occur when the user adds a new behaviour link between a protocol state machine and a class (*addBehaviour*), or when he links the message on a transition to an operation belonging to some class (*addReferredOp*), and so on.

In a similar way, we can answer question (2) by computing all sequential dependencies from a given primitive operation to all possible inconsistencies. This analysis results in the set of all possible, specified inconsistencies induced by the given primitive operation. As an example, the right part of Fig. 10 shows all sequential dependencies between the *addBehaviour* primitive operation and the inconsistency detection rules. The *addBehaviour* operation links a class to a statemachine. This operation can cause the occurrence of a *Dangling Operation Reference* and a *Abstract Statemachine* inconsistency. This set of inconsistency detection rules is complete with respect to the given primitive operation and the *specified* inconsistency detection rules.

6   Discussion and Future Research

Based on our experience with AGG's critical pair analysis, we can provide several recommendations. First, the algorithm sometimes reports false positives, which can be avoided by resorting to more clever solutions. Second, we found the analysis of transformation dependencies to be quite instrumental as a debugging mechanism, either to detect flaws in the type graph, or in the graph transformations themselves. We have encountered many such situations while carrying out the experiment reported on in this article.

Recently an integration of *AGG* within the Eclipse Modeling Framework (EMF) has been proposed [12]. Our work can directly benefit from this approach by integrating our ideas into the EMF environment. Once this will be achieved, we will be able to validate our approach on real industrial UML models. Such a validation will not significantly affect the results of this paper, but it will provide us with crucial information about the actual set of inconsistencies and resolution rules that are most frequently used in practice. Lange *et al.* have performed some empirical studies on the types of inconsistencies that commonly occur in industrial practice and how these can be resolved [13,14,15].

With respect to the issues of scalability, we are aware of the performance limitations of the AGG tool when applied to large models. While some of these limitations are inherent to the computational complexity of graph matching, we are convinced that the performance of the proposed approach can increase significantly, and can be made to scale up to large industrial projects, if an efficient implementation of a graph transformation tool were to be used. We have gathered some initial evidence to confirm our conviction. In [16] we performed a comparison of critical pair analysis in AGG and in a tool (called *Condor*) based on logic programming. We found that the latter approach resulted in a performance improvement of several orders of magnitude. A similar experiment was carried out by Hanh-Missi Tran (University of Lille, France) to compare AGG with logic programming approach for the purpose of evolving software architecture models. Again, a performance difference of several orders of magnitude was observed.

The fact that the resolution of one model inconsistency may introduce other inconsistencies is a clear sign of the fact that inconsistency resolution is a truly iterative and interactive process. One of the challenges is to find out whether the resolution process will ever *terminate*. It is easy to find situations that never terminate (cf. the presence of cycles in the dependency graph). Therefore, the challenge is to find out under which criteria a given set of resolution rules (for a given set of model inconsistencies and a given start graph) will terminate. Recent work that explores such termination criteria for model transformation based on the graph transformation formalism has been presented in [17].

Another challenge is to try and come up with an *optimal order* of resolution rules. For example, one strategy could be to follow a so-called "opportunistic resolution process", by always following the choice that corresponds to the least cognitive effort (i.e., the cognitive distance between the model before and after resolution should be as small as possible). How to translate this into more formal terms remains an open question. A second heuristic could be to avoid as much as possible resolution rules that give rise to *induced inconsistencies*.

We may also investigate research by the database community on automatic repairs on inconsistent databases [18]. This corresponds to a fully automated resolution process that intends to come to a final consistent model (by resolving all inconsistencies according to some predetermined formal strategy) that is as "semantically" close as possible to the original model. The strategy to be used depends on the precise notion of semantics that is used. A possible way to provide fully automated resolution of inconsistent models based on graph transformation may be achieved by relying on the *GROOVE* tool [19]. It can be used to explore the state space of all possible consistent models that can be obtained from an inconsistent one by applying a sequence of resolution rules.

Not all kinds of model inconsistencies and resolution rules can be expressed easily as graph transformation rules. For example, behavioural inconsistencies are also difficult to express in a graph-based way. Because of this, our tool has been developed in an extensible way, to make it easier to plug-in alternative mechanisms for detecting inconsistencies, such as those based on the formalism of description logics [20,21]. Of course, it remains to be seen how this formalism can be combined with the formalism of graph transformation, so that we can still benefit from the technique of critical pair analysis.

7   Related Work

Egyed [22] presents a very efficient approach to detect inconsistencies in UML models. This instant consistency checking mechanism scales up to large, industrial UML models. On the other hand, the inconsistency resolution process is not yet supported by this approach.

An approach that is very related to ours is reported in [23]. A logic rule-based approach (as opposed to a graph-based one) is proposed to detect and resolve inconsistencies in UML models, using the Java Rule Engine JESS. The architecture of this tool provides a Rule Engine Abstraction Layer, making it possible to replace their rule engine by a graph-based one.

The main novelty of our approach compared to the previously mentioned ones, is the use of the mechanism of critical pair analysis to detect mutual inconsistencies between rules that can be applied in parallel, as well as sequential dependency analysis between resolution rules.

There have been several attempts to use graph transformation in the context of inconsistency management. In [24], distributed graph transformation is used to deal with inconsistencies in requirements engineering. In [25], graph transformations are used to specify inconsistency detection rules. In [26] repair actions are also specified as graph transformation rules. Again, the added value of our approach is the ability to analyse conflicts and dependencies between detection and resolution rules.

8   Conclusion

In this article we addressed the problem of model inconsistency management. Resolution of inconsistencies occurring in or between models is supported in an iterative way by looking for model inconsistencies, and by proposing resolutions to remove these inconsistencies. Interactive tool support for this iterative inconsistency resolution process can benefit from a formal foundation. This article proposed a tool based on the underlying formalism of graph transformation. Given a formal specification of the UML model as a graph (and the metamodel as a type graph), model inconsistencies and their resolutions are specified as graph transformation rules. Furthermore, critical pair analysis is used to identify and analyse induced inconsistencies (i.e., new inconsistencies that are introduced after resolving existing inconsistencies), conflicting resolutions (i.e., applying a resolution invalidates other resolution rules) and cycles in the resolution process. Using critical pair analysis we can also compute all primitive operations that are needed to resolve a particular inconsistency, and all primitive operations that potentially induce a particular inconsistency. The results of these different analyses gives us the opportunity to reason about formal properties such as completeness, and allow us to improve tool support for the resolution process.

References

1. Balzer, R.: Tolerating inconsistency. In: Proc. Int'l Conf. Software Engineering, ACM (1991) 158–165
2. Object Management Group: Unified Modeling Language 2.0 Superstructure Specification. http://www.omg.org/cgi-bin/apps/doc?formal/05-07-04.pdf (2005)
3. Warny, J.F.: Détection et résolution des incohérences des modèles UML avec un outil de transformation de graphes. Master's thesis, Université de Mons-Hainaut, Belgium (2006)
4. Rozenberg, G., ed.: Handbook of graph grammars and computing by graph transformation: Foundations. Volume 1. World Scientific (1997)
5. Ehrig, H., Engels, G., Kreowski, H.J., Rozenberg, G., eds.: Handbook of graph grammars and computing by graph transformation: Applications, Languages and Tools. Volume 2. World Scientific (1999)
6. Plump, D.: Hypergraph rewriting: Critical pairs and undecidability of confluence. In: Term Graph Rewriting. Wiley (1993) 201–214

7. Taentzer, G.:   AGG: A graph transformation environment for modeling and validation of software. In: Proc. AGTIVE 2003. Volume 3062 of LNCS., Springer-Verlag (2004) 446–453
8. Ehrig, H., Prange, U., Taentzer, G.: Fundamental theory for typed attributed graph transformation. In: Proc. Int'l Conf. Graph Transformation. Volume 3256 of LNCS., Springer-Verlag (2004) 161–177
9. Hausmann, J.H., Heckel, R., Taentzer, G.: Detection of conflicting functional requirements in a use case-driven approach. In: Proc. Int'l Conf. Software Engineering, ACM (2002) 105–115
10. Mens, T., Taentzer, G., Runge, O.: Analyzing refactoring dependencies using graph transformation. Software and Systems Modeling (2007) To appear.
11. Mens, T., Van Der Straeten, R., D'Hondt, M.: Detecting and resolving model inconsistencies using transformation dependency analysis. In: Proc. MoDELS/UML 2006. Volume 4199 of LNCS., Springer-Verlag (2006) 200–214
12. Biermann, E., Ehrig, K., Köhler, C., Kuhns, G., Taentzer, G., Weiss, E.: Graphical definition of in-place transformations in the eclipse modeling framework. In: Proc. MoDELS/UML 2006. Volume 4199 of LNCS., Springer-Verlag (2006) 425–439
13. Lange, C.F., Chaudron, M.R.: An empirical assessment of completeness in UML designs. In: Proc. Int'l Conf. Empirical Assessment in Software Engineering. (2004) 111–121
14. Lange, C., Chaudron, M., Muskens, J.: In practice: UML software architecture and design description. IEEE Software 23 (2006) 40–46
15. Lange, C.F., Chaudron, M.R.: Effects of defects in UML models – an experimental investigation. In: Proc. Int'l Conf. Software Engineering, ACM (2006) 401–410
16. Mens, T., Kniesel, G., Runge, O.: Transformation dependency analysis - a comparison of two approaches. Série L'objet - logiciel, base de données, réseaux (2006) 167–182
17. Ehrig, H., Ehrig, K., de Lara, J., Taentzer, G., Varró, D., Varró-Gyapay, S.: Termination criteria for model transformation. In: Proc. Fundamental Aspects of Software Enginering (FASE). Volume 3442 of LNCS., Springer-Verlag (2005) 49–63
18. Wijsen, J.: Database repairing using updates. Trans. Database Systems 30 (2005) 722–768
19. Rensink, A.: The GROOVE simulator: A tool for state space generation. In: Proc. AGTIVE 2003. Volume 3062 of LNCS., Springer-Verlag (2004) 479–485
20. Van Der Straeten, R., Mens, T., Simmonds, J., Jonckers, V.: Using description logics to maintain consistency between UML models. In: UML 2003 - The Unified Modeling Language. Volume 2863 of LNCS., Springer-Verlag (2003) 326–340
21. Van Der Straeten, R.: Inconsistency Management in Model-driven Engineering. An Approach using Description Logics. PhD thesis, Department of Computer Science, Vrije Universiteit Brussel, Belgium (2005)
22. Egyed, A.: Instant consistency checking for the UML. In: Proc. Int'l Conf. Software Engineering, ACM (2006) 381–390
23. Liu, W., Easterbrook, S., Mylopoulos, J.: Rule-based detection of inconsistency in UML models. In: Proc. UML 2002 Workshop on Consistency Problems in UML-based Software Development, Blekinge Insitute of Technology (2002) 106–123
24. Goedicke, M., Meyer, T., , Taentzer, G.: Viewpoint-oriented software development by distributed graph transformation: Towards a basis for living with inconsistencies. In: Proc. Requirements Engineering 1999, IEEE Computer Society (1999) 92–99
25. Ehrig, H., Tsioalikis, A.: Consistency analysis of UML class and sequence diagrams using attributed graph grammars. In: ETAPS 2000 workshop on graph transformation systems. (2000) 77–86
26. Hausmann, J.H., Heckel, R., Sauer, S.: Extended model relations with graphical consistency conditions. In: Proc. UML 2002 Workshop on Consistency Problems in UML-Based Software Development. (2002) 61–74

Coalgebraic Modal Logic in CoCASL

Lutz Schröder and Till Mossakowski

Department of Computer Science, University of Bremen, and
DFKI Lab Bremen, Germany

Abstract. We propose to extend the algebraic-coalgebraic specification language CoCASL by full coalgebraic modal logic based on predicate liftings for functors. This logic is more general than the modal logic previously used in CoCASL and supports the specification of a variety of modal logics, such as graded modal logic, majority logic, and probabilistic modal logic. CoCASL thus becomes a modern modal language that covers a wide range of Kripke and non-Kripke semantics of modal logics via the coalgebraic interpretation.

Introduction

The algebraic-coalgebraic specification language CoCASL [14] combines the algebraic specification of functional aspects of software with the specification of reactive systems following the emerging coalgebraic paradigm [21]. As a specification logic, modal logic plays an analogous role for coalgebra as equational logic does for algebra; in particular, modal logic respects the behavioural encapsulation of the state space.

Two notions of modal logic have been included in the original design of Co-CASL. The first one treats observer operations that have a non-observable result sort as modalities (here, the sorts in the local environment are regarded as observable). The second notion of modal logic is based on a specific way of extracting modalities from datatypes with equations.

Recently, a more general formulation of *coalgebraic modal logic*, based on a notion of *predicate lifting*, has been proposed by Pattinson [18]. A substantial body of theory has developed for this generic logic, including results on duality, expressivity, decidability, and complexity [22, 9, 23, 26]. We hence propose to extend the CoCASL design by support for coalgebraic modal logic, including polyadic modal operators needed in order to obtain expressiveness in the general case, in particular for composite functors [22]. To this end, it is convenient to promote functors, which feature only implicitly in the original design of Co-CASL, to first-class citizens. Specifically, functors can be defined as algebraic or coalgebraic datatypes, with the dependency on type arguments recorded explicitly; such functors can then be used as signatures for coalgebraic types. Predicate liftings for given functors can be specified without polymorphic axioms using a classification result from [22]; one thus obtains customized modal logics for user-defined system types. In particular, it becomes possible to specify e.g. graded or probabilistic modal operators.

J.L. Fiadeiro and P.-Y. Schobbens (Eds.): WADT 2006, LNCS 4409, pp. 127–141, 2007.
© Springer-Verlag Berlin Heidelberg 2007

1   Coalgebraic Modal Logic

We briefly recapitulate the basics of the coalgebraic interpretation of modal logic.

Definition 1. [21] Let $T : \mathbf{Set} \to \mathbf{Set}$ be a functor, referred to as the *signature functor*, where $\mathbf{Set}$ is the category of sets. A *T-coalgebra* $A = (X, \xi)$ is a pair (X, ξ) where X is a set (of *states*) and $\xi : X \to TX$ is a function called the *transition* function. A *morphism* $(X_1, \xi_1) \to (X_2, \xi_2)$ of T-coalgebras is a map $f : X_1 \to X_2$ such that $\xi_2 \circ f = Tf \circ \xi_1$.

We view coalgebras as generalised transition systems: the transition function assigns to each state a structured set of successors and observations.

Coalgebraic modal logic in the form considered here has been introduced as a specification logic for coalgebraically modelled reactive systems in [18], generalising previous results [8, 20, 10, 17]. The coalgebraic semantics is based on predicate liftings; here, we consider the notion of polyadic predicate lifting introduced in [22].

Definition 2. An *n-ary predicate lifting* ($n \in \mathbb{N}$) for a functor T is a natural transformation

$$\lambda : \mathcal{Q}^n \to \mathcal{Q} \circ T^{\mathbf{op}},$$

where $\mathcal{Q}$ denotes the contravariant powerset functor $\mathbf{Set}^{\mathbf{op}} \to \mathbf{Set}$ (i.e. $\mathcal{Q}(f)(A) = f^{-1}[A]$), and $\mathcal{Q}^n$ refers to its n-fold cartesian product.

A coalgebraic semantics for a modal logic consists of a signature functor and an assignment of a predicate lifting to every modal operator; we write $[\lambda]$ for a modal operator that is interpreted using the lifting λ. Thus, a set Λ of predicate liftings for T determines the syntax of a modal logic $\mathcal{L}(\Lambda)$. Formulae $\phi, \psi \in \mathcal{L}(\Lambda)$ are defined by the grammar

$$\phi ::= \bot \mid \phi \wedge \psi \mid \neg \phi \mid [\lambda](\phi_1, \ldots, \phi_n),$$

where λ ranges over Λ and n is the arity of λ. Disjunctions $\phi \vee \psi$, truth $\top$, and other boolean operations are defined as usual.

The satisfaction relation $\models_C$ between states x of a T-coalgebra $C = (X, \xi)$ and $\mathcal{L}(\Lambda)$-formulae is defined inductively, with the usual clauses for the boolean operations. The clause for the modal operator $[\lambda]$ is

$$x \models_C [\lambda](\phi_1, \ldots, \phi_n) \iff \xi(x) \in \lambda(\llbracket \phi_1 \rrbracket_C, \ldots, \llbracket \phi_n \rrbracket_C)$$

where $\llbracket \phi \rrbracket_C = \{ x \in X \mid x \models_C \phi \}$. We drop the subscripts C when C is clear from the context.

Remark 3. Coalgebraic modal logic exhibits a number of pleasant properties and admits non-trivial metatheoretic results, e.g. the following.

1. States x and y in T-coalgebras A and B, respectively, are called *behaviourally equivalent* if there exists a coalgebra C and morphisms $f : A \to C$, $g : B \to C$ such that $f(x) = g(y)$. It is easy to see that coalgebraic modal logic is *adequate*, i.e. invariant under behavioural equivalence. Thus, coalgebraic modal logic automatically ensures encapsulation of the state space.

2. Conversely, it is shown in [18, 22] that if T is ω-accessible and Λ is separating in the sense that $t \in TX$ is determined by the set $\{(\lambda, A) \in \Lambda \times \mathcal{P}(X) \mid t \in \lambda(A)\}$, then $\mathcal{L}(\Lambda)$ is *expressive*, i.e. if two states satisfy the same $\mathcal{L}(\Lambda)$-formulae, then they are behaviourally equivalent.

3. As a consequence of 1., classes of coalgebras restricted by modal axioms have final models [11, 14].

4. The standard duality theory of modal logic, including the theory of ultra-filter extensions and bisimulation-somewhere-else, generalizes to coalgebraic modal logic [9].

5. Coalgebraic modal logic has the finite and shallow model properties [23, 26].

6. There are generic criteria for a coalgebraic modal logic to be (effectively) decidable [23, 26].

Coalgebraic modal logic subsumes a wide variety of modal logics (recall that a modal operator $\Box$ is called *monotone* if it satisfies $\Box(p \wedge q) \to \Box p$, and *normal* if it satisfies $(\Box p \wedge \Box q) \leftrightarrow \Box(p \wedge q)$).

Example 4. [18, 3, 23]

1. Let $\mathcal{P}$ be the covariant powerset functor. Then $\mathcal{P}$-coalgebras are graphs, thought of as transition systems or Kripke frames. The predicate lifting λ defined by

$$\lambda_X(A) = \{B \subset X \mid B \subset A\}$$

gives rise to the standard box modality $\Box = [\lambda]$. This translates verbatim to the finitely branching case, captured by the finite powerset functor $\mathcal{P}_{fin}$.

2. Coalgebras for the functor $N = \mathcal{Q} \circ \mathcal{Q}^{\mathrm{op}}$ (composition of the contravariant powerset functor with itself) are neighbourhood frames, the canonical semantic domain of non-normal logics [2]. The coalgebraic semantics induced by the predicate lifting λ defined by

$$\lambda_X(A) = \{\alpha \in N(X) \mid A \in \alpha\}$$

is just the neighbourhood semantics for $\Box = [\lambda]$.

3. Similarly, coalgebras for the subfunctor $\mathrm{Up}\mathcal{P}$ of N obtained by restricting N to upwards closed subsets of $\mathcal{Q}(X)$ are monotone neighbourhood frames [5]. Putting $\Box = [\lambda]$, with λ as above, gives the standard interpretation of the $\Box$-modality of monotone modal logic.

4. It is straightforward to extend a given coalgebraic modal logic for T with a set U of *propositional symbols*. This is captured by passing to the functor $T'X = TX \times \mathcal{P}(U)$ and extending the set of predicate liftings by the liftings λ^a, $a \in U$, defined by

$$\lambda_X^a(A) = \{(t, B) \in TX \times \mathcal{P}(U) \mid a \in B\}.$$

Since λ^a is independent of its argument, we can write the propositional symbol a in place of $[\lambda^a]\phi$, with the expected meaning.

5. The *finite multiset* (or *bag*) functor $\mathcal{B}_{\mathbb{N}}$ maps a set X to the set of maps $b : X \to \mathbb{N}$ with finite support. The action on morphisms $f : X \to Y$ is given by $\mathcal{B}_{\mathbb{N}} f : \mathcal{B}_{\mathbb{N}} X \to \mathcal{B}_{\mathbb{N}} Y, b \mapsto \lambda y. \sum_{f(x)=y} b(x)$. Coalgebras for $\mathcal{B}_{\mathbb{N}}$ are directed graphs with $\mathbb{N}$-weighted edges, often referred to as *multigraphs* [4], and provide a coalgebraic semantics for *graded modal logic* (GML): One defines a set of predicate liftings $\{\lambda^k \mid k \in \mathbb{N}\}$ by

$$\lambda_X^k(A) = \{b : X \to \mathbb{N} \in \mathcal{B}_{\mathbb{N}}(X) \mid \sum_{a \in A} b(a) > k\}.$$

The arising modal operators are precisely the modalities $\Diamond_k$ of GML [4], i.e. $x \vDash \Diamond_k \phi$ iff ϕ holds for more than k successor states of x, taking into account multiplicities. Note that $\Box_k$, defined as $\neg \Diamond_k \neg$, is monotone, but fails to be normal unless $k = 0$. A non-monotone variation of GML arises when negative multiplicities are admitted.

6. The *finite distribution functor* D_ω maps a set X to the set of probability distributions on X with finite support. Coalgebras for the functor $T = D_\omega \times \mathcal{P}(U)$, where U is a set of propositional symbols, are probabilistic transition systems (also called *probabilistic type spaces* [7]) with finite branching degree. The natural predicate liftings for T consists of the propositional symbols (Item 4 above) together with the liftings λ^p defined by

$$\lambda^p(A) = \{P \in D_\omega X \mid PA \geq p\}$$

where $p \in [0, 1] \cap \mathbb{Q}$. The induced operators are the modalities $L_p = [\lambda^p]$ of *probabilistic modal logic (PML)* [12, 7], where $L_p \phi$ reads 'ϕ holds in the next step with probability at least p'.

7. Let T be the functor given by $TX = \mathcal{Q}(X) \to \mathcal{P}(X)$ (with $\mathcal{P}$ the covariant powerset functor and $\mathcal{Q}$ the contravariant powerset functor). Then a T-coalgebra is a *standard conditional model* [2]. The strict implication operator $\Rightarrow$ of *conditional logic* is interpreted using the binary predicate lifting λ defined by

$$\lambda_X(A, B) = \{f : \mathcal{Q}(X) \to \mathcal{P}(X) \mid f(A) \subset B\}.$$

The following simple fact gives immediate access to all predicate liftings that a functor admits, and will serve as a means of specifying predicate liftings in CoCASL without introducing polymorphic axioms.

Proposition 5. *[22] For $n \in \mathbb{N}$, n-ary predicate liftings for T are in bijective correspondence with subsets of $T(2^n)$, where $2 = \{\top, \bot\}$. The correspondence works by taking a predicate lifting λ to $\lambda_{2^n}(\pi_1^{-1}\{\top\}, \ldots, \pi_n^{-1}\{\top\}) \subseteq T(2^n)$, where $\pi_i : 2^n \to 2$ is the i-th projection, and, conversely, $C \subseteq T(2^n)$ to the n-ary predicate lifting λ^C defined by*

$$\lambda_X^C(A_1, \ldots, A_n) = (T\langle \chi_{A_1}, \ldots, \chi_{A_n} \rangle)^{-1}[C]$$

for $A_i \subseteq X$ ($i = 1, \ldots, n$), where angle brackets denote tupling of functions and $\chi_A : X \to 2$ is the characteristic function of $A \subseteq X$.

(We refrain from defining the logic using subsets of $T(2^n)$ instead of predicate liftings, as the latter convey a better intuition of how a given modal operator is interpreted.)

2   Algebraic-Coalgebraic Specification

The algebraic-coalgebraic specification language CoCASL has been introduced in [14] as an extension of the standard algebraic specification language CASL. For the basic CASL syntax, the reader is referred to [1, 15]. We briefly explain the CoCASL features relevant for the understanding of the present work using the example specification shown in Fig. 1.

---

spec UNIT =
  sort *Unit*
  • $\forall x, y : Unit . x = y$

spec STREAM [**sort** *Elem*] **given** UNIT =
  cotype *Stream* ::= (*hd* :? *Elem*; *tl* :? *Stream*) | (*stop* :? *Unit*)

spec FAIRSTREAM [**sort** *Elem*] **given** UNIT =
  op *c* : *Elem*
  then cofree { STREAM [**sort** *Elem*] **with** *Stream* $\mapsto$ *FairStream*
      then
      • $\langle tl* \rangle hd = c$
  }
end

Fig. 1. Specification of a fairness property

Dually to CASL's datatype construct **type**, CoCASL offers a **cotype** construct which defines coalgebraic process types; it is formally proved in [14] that one can indeed define for each cotype signature a functor T such that models of the cotype correspond to T-coalgebras. A simple example is the cotype of fair streams defined in Fig. 1. We first introduce a singleton type *Unit*. Then, a loose type of possibly terminating streams over a (loosely interpreted) sort *Elem* is declared using a cotype with two alternatives. Like a type declaration, a cotype declaration is essentially just a short way of declaring operations; specifically, the declaration of *Stream* produces *observer* operations $hd : Stream \rightarrow? Elem$, $tl : Stream \rightarrow? Stream$, and $stop : Stream \rightarrow? Unit$, with additional conditions on the definedness of observers which guarantee that models of the cotype *Stream* are essentially coalgebras for the functor $\lambda X. Elem \times X + 1$ (i.e. deterministic output automata with output in *Elem*).

CoCASL's modal logic now turns the observers hd and $stop$ into flexible constants, as they have result sorts which stem from the local environment and hence are regarded as observable, while the observer tl has a result sort which is regarded as non-observable and hence induces modalities $[tl]$ and $\langle tl \rangle$. The

modality [*tl*] is interpreted as 'for the tail of the stream, if any, it is the case that
...', while ⟨*tl*⟩ means 'the stream has a tail, which satisfies ...'. The latter is
stronger, as it enforces that the tail is defined.

Modalities can be starred; this refers to states reachable by a finite number
of iterated applications of an observer. The specification FAIRSTREAM of Fig. 1
hence expresses that all fair streams will always eventually output a *c* (before
they possibly end). Here, the 'always' stems from the fact that by stating the
modal formula, we mean that it holds for all elements of type *FairStream*.

The keyword **cofree** in the specification FAIRSTREAM further restricts the
models to those that are final (over their *Elem*-part). In particular, one has a
coinduction principle for *FairStream*, and all possible behaviours are realised
by the cotype *FairStream* — i.e. up to isomorphism, the cotype *FairStream*
consists of all streams satisfying the modal axiom.

This form of specification only supports (coalgebras for) polynomial functors.
A more general form of coalgebras involves *structured observations*: e.g. for non-
deterministic automata, in each state, a *set* of successor states can be observed.
In Fig. 2, finite sets are specified using a *free* type. Note that without structured
observers, there can only be one or no successor state for a given state, while
now we have a finite set of successor states. Hence, the modal logic needs to
be adapted accordingly. As before, each observer with a (possibly structured)
non-observable result leads to a modality. Since the observer *next* is additionally
parameterized over an input sort *In*, we have modalities [*next(i)*] and ⟨*next(i)*⟩
for *i* : *In*. The interpretation of these modalities is 'after reading *i* in the current
state, for *all* successor states, it is the case that ...' and 'after reading *i* in the
current state, for *some* successor state, it is the case that ...', respectively. E.g.
the specification of non-deterministic automata in Fig. 2 uses these operators to
express a form of liveness property stating that if an input *i* is disabled in some
state, then there exists a sequence of *tau*-transitions that will enable *i*.

```
spec LIVENONDETERMINISTICAUTOMATA =
    sort In
    op  tau : In
    sort State
    then  free %modal{
        type  Set ::= {} | {_}(State) | _∪_(Set; Set)
        op  _∪_ : Set × Set → Set,
                assoc, comm, idem, unit {}   }
    then  cotype State ::= (next  : In → Set)
    •  ∀i : In • [next(i)]false ⇒ ⟨next(tau)*⟩⟨next(i)⟩ true
end
```

Fig. 2. Specification of 'live' non-deterministic automata using modalities for structured observations

The example can be recast in the framework of section 1 as follows. The annotation %**modal** leads to extraction of the finite powerset functor $\mathcal{P}_\omega$, and the cotype leads to a functor

$$T X = \mathcal{P}_\omega(X)^{In}$$

On this functor, a canonical predicate lifting is induced via

$$nat(\mathcal{Q}, \mathcal{Q} \circ T^{\mathbf{op}}) \cong \mathcal{Q}(T2) \ni T\{\top\}.$$

The modal logic for structured observations described above is just the modal logic induced by this predicate lifting.

While this form of modal logic for CoCASL is syntactically rather lightweight, it leads to the need of carrying around distinguished presentations (constructors and equations) of datatypes in the signatures (and these presentations need to be preserved by signature morphisms). Moreover, the interaction between basic and structured specifications indicated by the annotation %**modal** is rather implicit and hard to grasp. Most severely, the approach can only handle specific predicate liftings, and hence has only limited expressiveness.

3   Functors and Liftings in CoCASL

Motivated by the above considerations, we extend CoCASL by explicit notions of functors and modalities. Both these concepts will, in the extended language, give rise to named components of signatures.

Like in modern higher order functional programming languages such as Haskell [19], functors constitute *type constructors* that enrich the type system generated by the signature. The semantics of functors requires that these type constructors are really the object parts of functors, although the action of the functor on maps is not directly syntactically available as higher order functions are not a basic CoCASL language feature (function types and higher order functions may however by specified by the user; cf. [14]).

A functor is introduced by the keyword **functor**. It must be *defined* as either an initial datatype or a final process type. Thus, the definition takes one of the two standard forms

functor $F(X) = $ **free** { **type** $F(X) ::= \ldots$ }

or

functor $F(X) = $ **cofree** { **cotype** $F(X) ::= \ldots$ }

where the omitted parts consist of a single type or cotype declaration, respectively, optionally followed by declarations of additional operation and predicate symbols, as well as

- in the case of a free type declaration, Horn axioms constraining the type as well as the additional operations (typically providing recursive definitions for the latter)
- in the case of a cofree cotype declaration, modal axioms over the cotype $F(X)$ and corecursive definitions of the additional operations and predicates.

Note that the use of modal axioms in the second case does not constitute a circularity: $F(X)$ is defined as the final coalgebra for a previously declared functor, for which modalities have already been defined. Additional operations and predicates introduced along with the definition of the functor are uniquely defined due to the freeness or cofreeness constraint, respectively.

The format of functor definitions has been chosen in such a way that, given any interpretation of the functor argument X as a set A, the type $F(X)$ has an interpretation which is determined uniquely up to isomorphism and depends functorially on A; thus, F *induces* an endofunctor $[\![F]\!]$ on the category of sets. In the case of a free datatype, this functor takes a set A to the interpretation of $F(X)$ in the initial model of the defining specification interpreting X as A, and correspondingly with initial models replaced by fibre-final models in the case of a cofree cotype (cf. [14] for the definition of fibre finality).

At the level of the static semantics, functor definitions such as the above have the effect of extending the signature by a type constructor (F in the above examples), to which the corresponding basic specification in curly brackets is explicitly associated. The latter, in turn, has an enlarged local environment where the functor argument (X in the above examples) appears as an additional sort symbol; of course, the functor argument is hidden in the subsequent specification.

In the model semantics, the functor definition as such does not have any effect at all — there is no need for recording an explicit interpretation of the functor in the models, as the interpretation is already determined (up to isomorphism) by the remaining parts of the model. This interpretation shows up, however, as soon as the functor is actually used. As functors are regarded as type constructors, we have a type formation rule producing for every type s and every functor F a type $F(s)$, whose interpretation is obtained from the interpretation of s by applying the functor $[\![F]\!]$ induced by F as described above. A typical use of functors is in types of observers for cotypes; in particular, a coalgebra X for a functor F is declared by writing

cotype $X ::= (next : F(X))$

The second place where functors may appear is in definitions of predicate liftings. Unary predicate liftings are introduced by means of the keyword **modality** in the form

modality $m : F = \{C \bullet \phi\}$

This declares m to be a unary predicate lifting for F, defined as corresponding to the subset $\{C \mid \phi\}$ of $F2$ under Prop. 5. Polyadic modalities may be declared in the form

n placeholders

modality $m(\overbrace{\_\_;\ldots;\_\_}) : F = \{C \bullet \phi\}$

The above declares m to be an n-ary predicate lifting for F, corresponding under Prop. 5 to the subset $\{C \mid \phi\}$ of $F(2^n)$.

There is a certain amount of additional syntax available for purposes of defining the properties ϕ above. To begin, the local variable mentioned by ϕ (C in the above example) need not (and in fact cannot) be provided with a type, being implicitly of type $F(2^n)$; moreover, within the scope of ϕ, further variables without explicit typing may be used (in quantifications) that represent values of type $2 = \{\bot, \top\}$. Finally, the elements of 2 may be explicitly referred to as terms *true* and *false* (in standard CASL, *true* and *false* are *formulas*). All this serves to encapsulate the mention of 2 within definitions of liftings, rather than introducing a type of truth values globally, in an effort to keep the language extension as non-invasive as possible.

CoCASL's original implicit mechanism for defining modalities is kept in the extension: by writing

modality $m : F$ **canonical**

m is defined to be the predicate lifting for F corresponding to the subset $F\{\top\}$ of $F2$.

The modal operator induced by a predicate lifting m for T and an observer operation $f : X \to TX$ of a cotype is standardly denoted as $[m; f]$. As this is frequently not the desired notation in particular cases, we provide syntax annotations that allow replacing the standard syntax with rather arbitrary notation. Explicitly, the annotation

modality $m(\_\_;\ldots;\_\_) : F = \{C \bullet \phi\}$ **%syntax** $[m; \_\_;\ldots;\_\_] = L$

introduces the mixfix identifier L, containing n placeholders corresponding to the arguments of $[m; \_\_;\ldots;\_\_]$ in the given order, as an alternative notation for the n-ary modal operator $[m; \_\_;\ldots;\_\_]$. (Of course, the annotation may be used with **canonical** as well.)

In this general setting, iterated modal operators $[m; f*]$ are defined as greatest fixed points $[m; f*]\phi = \nu X. \phi \wedge [m; f]X$. More precisely, the semantics is defined as the union of all fixed points, which yields a greatest fixed point if the modal operator $[m; f]$ is monotone. The dual operator is defined by $\langle m; f* \rangle \equiv \neg[m; f*]\neg$.

Remark 6. A basic motivation for introducing functors and modalities as dedicated language features is to avoid the need for extending the language by shallow polymorphism as e.g. in the higher order CASL extension HASCASL [25]. The crucial point here is that polymorphic *axioms* complicate the semantics [24]. It is for this reason that functors and predicate liftings are provided with mandatory definitions, since loosely specified functors or liftings would give rise to implicit polymorphic functoriality or naturality axioms, respectively. Similarly, defining predicate liftings via the correspondence of Prop. 5 serves the purpose of avoiding polymorphic definitions. Finally, the possibility of including auxiliary operations and predicates in the format for functor definitions provides a workaround replacing later polymorphic definitions of such entities. If the later introduction of polymorphic operations or predicates is desired by the user, e.g. for purposes of specification structuring, then these may (resp. have to) be emulated by means of parametrized specifications.

Remark 7. Like the original version of CoCASL's modal logic, functors and modal operators induced by predicate liftings may be regarded as a syntactic sugaring of basic CoCASL. Functors may be replaced by parametrized specifications of the associated type constructor and the action on morphisms (which has two formal type parameters), and their application by explicit instantiations. In order to code the modal operator $[m; f]$ induced by a predicate lifting m for T and an observer $f : X \to TX$ of a cotype X (the general case of mutually recursive cotypes (X_i) with observers $f : X_i \to TX_j$ and polyadic liftings works analogously), one specifies 2 as a free datatype and the predicate type $\mathcal{Q}(X)$ as a cofree cotype with observer $is\_in : \mathcal{Q}(X) \times X \to 2$, correspondingly for the predicate type $\mathcal{Q}(T2)$. The definition of m then induces an element of $\mathcal{Q}(T2)$.

Remark 8. The mechanism for defining functors described above is, as the examples given in the next section will show, quite flexible. It does have its limitations, however; in particular, it does not cover definitions of functors **Set** $\to$ **Set** that come about as composites of functors involving a third category, notably composites of the form **Set** $\to$ **Set**$^{\mathbf{op}}$ $\to$ **Set**. This includes e.g. the neighbourhood frame functor and the standard conditional model functor (Examples 4.2 and 4.7). A more general mechanism for functor definitions would include explicit definitions of the action of the functor on morphisms, something we are trying to avoid for reasons given in Remark 6.

4   Example Specifications

We now illustrate the specification of modal logics in CoCASL by means of a number of examples, some of them formal specifications of logics given in Example 4.

To begin, Fig. 3 shows a specification of standard modal logic, interpreted over finitely branching Kripke frames. The latter are specified as coalgebras for the finite powerset functor, called Set in the specification. Recursive functions and predicates needed at later points are defined along with the recursive datatype $Set(X)$ itself; in particular, the subset predicate is needed for the definition of the modality all in Fig. 3, and the elementhood predicate is provided for later use in Fig. 5. The modality all is the box modality of standard modal logic; it arises from the predicate lifting taking a set $A \subseteq X$ to the set $Set(A) \subset Set(X)$, which corresponds according to Prop. 5 to the subset $\{\emptyset, \{\top\}\}$ of $Set(2)$.

spec FINITEBRANCHING =
functor $Set(X) =$ **free** {
  type $Set(X) ::= \{\} \mid \{\_\}(X)$
  op $\_ \cup \_ : Set(X) \times Set(X) \to Set(X),$ **assoc, comm, idem, unit** $\{\}$
  preds $\_ \subseteq \_ : Set(X) \times Set(X)$
        $\_ \epsilon \_ : X \times Set(X)$
      $\ldots$ %% inductive definitions of $\subseteq$, ϵ
  }
modality $all : Set = \{C \bullet C \subseteq \{true\}\};$ %**syntax** $[all; \_] = [\_]$

Fig. 3. Specification of finitely branching modal logic

The modality *all* may alternatively be defined as a canonical modality corresponding to the subset $Set(\{\top\}) = \{\emptyset, \{\top\}\}$ of $Set(2)$:

modality $all : Set$ **canonical** %**syntax** $[all; \_] = [\_]$

It may be used in specifications such as

free type $Bit ::= 0 \mid 1$
cotype $Node ::= (next : Set(Node); out : Bit)$
  $\bullet$ $out = 0 \Rightarrow [next]\, out = 1$

(with explicit mention of *all* suppressed as prescribed by the syntax annotation) declaring a (loose) transition system with bit-labelled nodes, where all the successors of nodes labelled 0 are labelled 1.

Using cofree cotypes, one can also specify the full class of Kripke frames, without a bound on branching, as the semantics of standard modal logic; the corresponding specification in shown in Fig. 4. Here, the functor *Set* denotes the full powerset functor, specified as a cofree cotype observed via a boolean elementhood function. Note that generally, observer declarations with functional result type actually declare an uncurried function, in this case $\_ \epsilon \_ : Set(X) \times X \to Bool$.

spec UNBOUNDEDBRANCHING =
free type $Bool ::= F \mid T$
functor $Set(X) =$ **cofree** {**cotype** $Set(X) ::= (\_ \epsilon \_ : X \to Bool)$ }
modality $all : Set = \{C \bullet false\ \epsilon\ C = F\};$ %**syntax** $[all; \_] = [\_]$

Fig. 4. Specification of modal logic with unbounded branching

As an example of a binary modality, a standard modality for the composite of the finite powerset functor and the squaring functor is specified in Fig. 5. This functor is well suited for the modelling of processes that may fork into independent subprocesses, e.g. for purposes of higher order communication [27] or mobility [6]; it does not admit an expressive set of unary modalities [22], so that the use of a binary modality cannot in general be avoided. The specification in Fig. 5 imports the definition of the finite powerset functor from Fig. 3; the composition of this functor with the squaring functor $Pair$ is realized by means of a free datatype $PairSet(X)$ encapsulating the type $Set(Pair(X))$. The binary modality $bAll$ is determined by the binary predicate lifting taking a pair (A, B) of subsets of X to the subset $\{C \mid (a, b) \in C \implies a \in A \wedge b \in B\}$ of $PairSet(X)$, which corresponds under Prop. 5 to the subset $\{C \mid ((a, b), (c, d)) \in C \implies a = \top \wedge d = \top\}$ of $PairSet(2^2)$ (where we identify 2^2 with 2×2). The modality $bAll$ can be used in specifications such as

free type $Bit ::= 0 \mid 1$
cotype $Proc ::= (branch : PairSet(Proc); out : Bit)$
• $out = 0 \Rightarrow [branch](out = 1, out = 0)$

declaring a (loose) cotype $Proc$ of forking processes with bit-labelled states, where all left (right) children of states labelled 0 are labelled 1 (0).

---

spec FORK = FINITEBRANCHING **then**
functor $Pair(X) =$ **free** $\{$**type** $Pair(X) ::= pair(X; X)\}$
functor $PairSet(X) =$ **free** $\{$
   type $PairSet(X) ::= pairSet(Set(Pair(X)))$
   pred $\_ \epsilon \_ : Pair(X) \times PairSet(X)$
   vars $z : Pair(X); A : Set(Pair(X))$
   • $z \epsilon pairSet(A) \Leftrightarrow z \epsilon A$
$\}$
modality $bAll(\_; \_) : PairSet = \{C \bullet \forall a, b, c, d \bullet$
          $pair((a, b), (c, d)) \epsilon C \Rightarrow a = true \wedge d = true\}$
                           %**syntax** $[bAll; \_] = [\_]$

Fig. 5. A binary modality for the 'forking functor' $\lambda X. \mathcal{P}(X \times X)$

As an illustration of a coalgebraic semantics which is more clearly distinct from standard Kripke semantics, a specification of graded modal logic, interpreted over coalgebras for the bag functor, is shown in Fig. 6. The bag functor is obtained, like the finite powerset functor, from basic operations representing empty, singletons, and union, without however imposing idempotence on the union operator. The modal operators of GML are induced by predicate liftings λ^k as in Example 4.5. The correspondence of Prop. 5 takes λ^k to the subset $\{n\top + m\bot \mid n > k\}$ of $Bag(2)$.

The following example specification of a bag-branching bit-labelled process type expresses that there are always more successors labelled 1 than successors labelled 0:

free type $Bit ::= 0 \mid 1$
cotype $M ::= (next : Bag(M); out : Bit)$
var $n : Nat$
• $\langle n, next \rangle out = 0 \Rightarrow \langle n + 1, next \rangle out = 1$

Note how quantification over indices of modal operators increases expressivity; indeed, the above formula corresponds to the formula $M(out = 1)$ of *majority logic* [16], which is not standardly expressible in graded modal logic.

---

spec GRADEDMODALLOGIC =
functor $Bag(X) =$ **free** {
  type $Bag(X) ::= \{\} \mid \{\_\}(X)$
  op $\_ \cup \_ : Bag(X) \times Bag(X) \to Bag(X),$
        assoc, comm, unit $\{\}$
  op $count : Bag(X) \times X \to Nat$
  ...   %% recursive definition of *count*
  }
var $n : Nat$
modality $more(n) : Bag = \{C \bullet count(C, true) > n\}$
           %**syntax** $[more(n); \_] = \langle n; \_ \rangle$

Fig. 6. Specification of Graded Modal Logic

We conclude with a few examples illustrating the interpretation of starred modalities:

— Over the cotype *Node* specified above, $[next*]$ corresponds to the CTL operator AG, i.e. $[next*]\phi$ holds in a state if all states reachable from it in finitely many steps satisfy ϕ. The dual $\langle next* \rangle$ of $[next*]$ corresponds to the CTL operator EF.
— If we introduce a separate diamond operator for *Set*

  modality $ex : Set = \{C \bullet \neg(false \in C)\}$

  then $[ex; next]$ corresponds to the CTL operator EG, with dual AF.
— Over the bag-branching cotype M defined above, $[more(2)*]\phi$ is satisfied in a state $x : M$ iff ϕ holds everywhere on some bag-branching substructure of M with root x in which every node has at least 2 children, counting multiplicities.

5 Conclusion

We have proposed a syntactic integration of recent forms of coalgebraic modal logic into CoCASL. The main device is a syntax that makes functors and predicate liftings explicit, replacing less flexible implicit mechanisms in the original CoCASL design. This leads to both a cleaner (static) semantics of CoCASL specifications and to increased expressiveness: in the resulting coalgebraic modal logic, one can now express graded modal logic, majority logic, and probabilistic modal logic, as well as binary modalities. We have illustrated these concepts by means of extensive example specifications.

Once these extensions are incorporated into the CoCASL tool support, coalgebraic modal logic will be embedded into an extensive network of related specification languages and tools, in particular theorem provers, within the Bremen heterogeneous tool set Hets [13]. This will also provide a suitable framework for experimental implementations of generic decision procedures for coalgebraic modal logic [26].

References

[1] M. Bidoit and P. D. Mosses. CASL *User Manual*, volume 2900 of *LNCS*. Springer, 2004.

[2] B. Chellas. *Modal Logic*. Cambridge, 1980.

[3] C. Cîrstea and D. Pattinson. Modular construction of modal logics. In *Concurrency Theory, CONCUR 04*, volume 3170 of *LNCS*, pages 258–275. Springer, 2004.

[4] G. D'Agostino and A. Visser. Finality regained: A coalgebraic study of Scott-sets and multisets. *Arch. Math. Logic*, 41:267–298, 2002.

[5] H. H. Hansen and C. Kupke. A coalgebraic perspective on monotone modal logic. In *Coalgebraic Methods in Computer Science, CMCS 04*, volume 106 of *ENTCS*, pages 121–143. Elsevier, 2004.

[6] D. Hausmann, T. Mossakowski, and L. Schröder. A coalgebraic approach to the semantics of the ambient calculus. *Theoret. Comput. Sci.*, 366:121–143, 2006.

[7] A. Heifetz and P. Mongin. Probabilistic logic for type spaces. *Games and Economic Behavior*, 35:31–53, 2001.

[8] B. Jacobs. Towards a duality result in the modal logic of coalgebras. In *Coalgebraic Methods in Computer Science, CMCS 00*, volume 33 of *ENTCS*. Elsevier, 2000.

[9] C. Kupke, A. Kurz, and D. Pattinson. Ultrafilter extensions for coalgebras. In *Algebra and Coalgebra in Computer Science, CALCO 05*, volume 3629 of *LNCS*, pages 263–277. Springer, 2005.

[10] A. Kurz. Specifying coalgebras with modal logic. *Theoret. Comput. Sci.*, 260:119–138, 2001.

[11] A. Kurz. Logics admitting final semantics. In *Foundations of Software Science and Computation Structures, FOSSACS 02*, volume 2303 of *LNCS*, pages 238–249. Springer, 2002.

[12] K. Larsen and A. Skou. Bisimulation through probabilistic testing. *Inform. Comput.*, 94:1–28, 1991.

[13] T. Mossakowski. Heterogeneous specification and the heterogeneous tool set. Habilitation thesis, University of Bremen, 2004.

[14] T. Mossakowski, L. Schröder, M. Roggenbach, and H. Reichel. Algebraic-co-algebraic specification in CoCASL. *J. Logic Algebraic Programming*, 67:146–197, 2006.

[15] P. D. Mosses, editor. CASL *Reference Manual*, volume 2960 of *LNCS*. Springer, 2004.

[16] E. Pacuit and S. Salame. Majority logic. In *Principles of Knowledge Representation and Reasoning, KR 04*, pages 598–604. AAAI Press, 2004.

[17] D. Pattinson. Semantical principles in the modal logic of coalgebras. In *Theoretical Aspects of Computer Science, STACS 01*, volume 2010 of *LNCS*, pages 514–526. Springer, 2001.

[18] D. Pattinson. Expressive logics for coalgebras via terminal sequence induction. *Notre Dame J. Formal Logic*, 45:19–33, 2004.

[19] S. Peyton-Jones, editor. *Haskell 98 Language and Libraries — The Revised Report*. Cambridge, 2003. Also: J. Funct. Programming **13** (2003).

[20] M. Rößiger. Coalgebras and modal logic. In *Coalgebraic Methods in Computer Science, CMCS 00*, volume 33 of *ENTCS*. Elsevier, 2000.

[21] J. Rutten. Universal coalgebra: A theory of systems. *Theoret. Comput. Sci.*, 249:3–80, 2000.

[22] L. Schröder. Expressivity of coalgebraic modal logic: the limits and beyond. In *Foundations of Software Science and Computation Structures, FOSSACS 05*, volume 3441 of *LNCS*, pages 440–454. Springer, 2005. Extended version to appear in *Theoret. Comput. Sci.*

[23] L. Schröder. A finite model construction for coalgebraic modal logic. In L. Aceto and A. Ingólfsdóttir, editors, *Foundations of Software Science and Computation Structures, FOSSACS 06*, volume 3921 of *LNCS*, pages 157–171. Springer, 2006. Extended version to appear in *J. Logic Algebraic Programming*.

[24] L. Schröder, T. Mossakowski, and C. Lüth. Type class polymorphism in an institutional framework. In *Recent Developments in Algebraic Development Techniques, 17th International Workshop, WADT 04*, volume 3423 of *LNCS*, pages 234–248. Springer, 2005.

[25] L. Schröder, T. Mossakowski, and C. Maeder. HASCASL – Integrated functional specification and programming. Language summary. Available at http://www.informatik.uni-bremen.de/agbkb/forschung/formal_methods/CoFI/HasCASL, 2003.

[26] L. Schröder and D. Pattinson. PSPACE reasoning for rank-1 modal logics. In *Logic in Computer Science, LICS 06*, pages 231–240. IEEE, 2006.

[27] B. Thomsen. A theory of higher order communicating systems. *Inform. and Comput.*, 116:38–57, 1995.

SV$_t$L: System Verification Through Logic Tool Support for Verifying Sliced Hierarchical Statecharts

Sara Van Langenhove* and Albert Hoogewijs

Department of Pure Mathematics and Computer Algebra
Ghent University, Belgium
{Sara.VanLangenhove,Albert.Hoogewijs}@UGent.be

Abstract. SV$_t$L is the core of a slicing-based verification environment for UML statechart models. We present an overview of the SV$_t$L software architecture. Special attention is paid to the slicing approach. Slicing reduces the complexity of the verification approach, based on removing pieces of the model that are not of interest during verification. In [18] a slicing algorithm has been proposed for statecharts, but it was not able to handle orthogonal regions efficiently. We optimize this algorithm by removing false dependencies, relying on the broadcasting mechanism between different parts of the statechart model.

1 Introduction

The industry strives for a reliable, high-quality and time-efficient UML design methodology. Formal verification methods support the design of a wide-range of systems and advertise these benefits, yet the industry remains skeptical [4, 7]; verifying systems is still done by experts with limited support from automated verification tools. That makes the verification process often error-prone.

The SV_tL (System Verification through Logic) framework addresses these problems and supports a "provably correct design" methodology. The framework enhances the commercial UML tool Rhapsody, by adding a slicing-based verification functionality. This approach to verification advocates software designers to apply formal methods for improving the quality of systems whose behavior is designed using UML statechart models. To be of interest in practice, the tool fulfills the following requirements:

- The framework provides tool support for verification-in-the-large. It allows both a thorough verification of systems that are modelled as multi-threaded applications and huge statechart designs as it integrates slicing as an abstraction (reduction) approach.
- The tool bridges the semantic gap between statecharts and verification logic; and thus hides away "scary" logical aspects.

* Funded by Ghent University (BOF/GOA project B/03667/01 IV1).

J.L. Fiadeiro and P.-Y. Schobbens (Eds.): WADT 2006, LNCS 4409, pp. 142–155, 2007.

– It delivers interaction and assurance in a form fit for designers. Designers
use the tool without a training effort in the field of formal verification.

SV$_t$L bridges the gap between Rhapsody and the model checker Cadence SMV
(CaSMV [12]) as Fig. 1 illustrates. Developers use *Rhapsody in J*[1] to define the
(multi-threaded) behavior of systems as a set of UML statechart diagrams. The
Property Writing Assistant offers specification patterns [3] to guide developers in
defining system requirements. The framework imports the model and performs
the necessary transformations and simplifications (*Parser, Slicer, Generator*) to
retrieve a semantically equivalent verification model. It also translates the requirements into CaSMV temporal logic formulas. The tool verifies the properties
by spawning the model checker as an external process. If the temporal formula
fails to hold, the *GUI* visualizes the verification failure in the context of UML
statechart diagrams since the error traces on the CaSMV level are hardly readable for the designers working at the UML level. Designers either change the
requirement or the behavioral model and resume the verification.

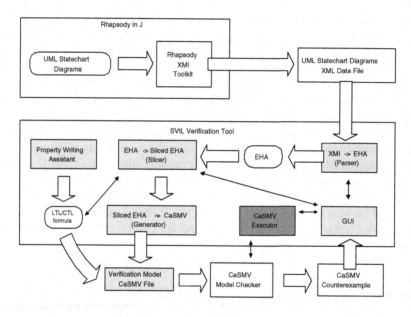

Fig. 1. Functionality of SV$_t$L

Outline of the paper. In the next section we describe a multi-threaded behavioral
model, which will be used to explain the slicing principle of SV$_t$L. Section 3
introduces two slicing algorithms that are explored in the sections thereafter.
Section 4 emphasizes on slicing a single statechart. The definitions presented
are concentrating on the given example (for the full description we refer to [10]).

---

[1] Available from http://www.ilogix.com/homepage.aspx.

Section 5 elaborates the slicing principles for a multi-threaded behavioral model. They are based on an extension of the happens-before relation to collaborating statecharts. Section 6 considers the thread-run-to-completion semantics as underlying semantics for the verification model of the multi-threaded behavioral model. Section 7 shortly refers to the equivalences of the models used in the verification process.

Related Tools. There are several tools for automated verification of UML statechart models [5, 14, 16], but they place restrictions on the characteristics of the multi-threaded behavioral model. The most innovative characteristic of the SV_tL framework is that it is the first verification tool capable to handle huge designs due to the integration of a slicing approach. Therefore the slicing-based verification technology is considered to be the strength of SV_tL.

2   The Multi-threaded Behavioral Model

In this paper we propose a new slicing technique in the context of the verification of concurrent multi-threaded models. In general, the problem is known to be undecidable [17]. Based on communication using locks, Kahlon and Gupta [8] show that model checking threads for LTL properties becomes feasible. Our slicing technique proceeds along similar lines.

A simple example of a multi-threaded behavioral model comes in a washing machine. As the machine runs through its washing cycle, the android Marvin[2] carries out other tasks in the household. In terms of UML, both the washing machine and Marvin are interacting *active (concurrent) objects*. The washing machine contains a water tank and an electrical heater. These components are *passive (sequential) objects* since they execute one after the other. Due to the presence of active objects, we acquire a multi-threaded design. Each tread of control contains at most one active object, and allows to include an arbitrary number of passive objects in the group (Fig. 2). SV_tL allows that both active and passive ("sequential") objects of the model can be reactive, i.e. statechart diagrams specify their behaviors. Verification tools like [5, 14, 16] only allow active objects to be reactive which definitely restricts the complexity of a realistic design.

Let us zoom in on a simplified statechart diagram of the washing machine depicted in Fig. 3. Each washing cycle consists of several activities as visualized in the concurrent region WashingCycle. Marvin can configure the program mode (behavior of the concurrent region Mode) at any time prior to the start of a wash cycle. The program mode dictates the wash, rinse and dry times. At the moment Marvin starts a wash cycle (transition t1), a lamp is blinking. The lamp stops blinking when the wash is done (behavior depicted in region Lamp). Note that Marvin is also responsible for refilling the soap tank. The event evSoapEmpty (entry action of state GetSoap) triggers a signal Marvin can react on whenever it is convenient.

---

[2] Marvin is the paranoid android of the "The Hitchhiker's Guide to the Galaxy".

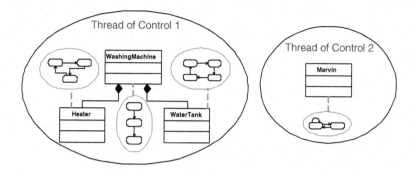

Fig. 2. Example of a Multi-Threaded Behavioral Model

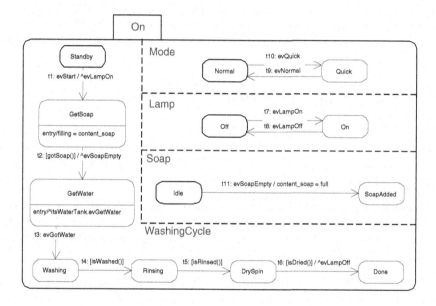

Fig. 3. Statechart of the Washing Machine

Figure 4 shows the behavior of the water tank. This component is activated by the washing machine by sending out the event **evGetWater** (entry action of state **GetWater**). The water tank sends the event **evGotWater** back to the washing machine when the water level has reached a certain limit.

3 The Power of the Tool: Slicing

Model checking is only feasible if the input model is small; the smaller the better. Slicing simplifies the verification of behavioral models with respect to a property of interest. It reduces the input model while keeping the relevant

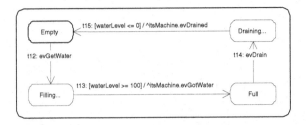

Fig. 4. Statechart of the Water Tank

elements for verification at hand. As such SV_tL is capable to verify larger UML statechart input models. SV_tL exploits two slicing algorithms: an algorithm to slice a single statechart (e.g. the statechart of Fig. 3), and a parallel algorithm to slice complex multi-threaded models (e.g. the model of Fig. 2). The verification property is satisfied by the reduced behavioral model if and only if it is satisfied by the full model [18, 10].

4   Slicing a Single Statechart

The slicer of SV_tL is based (i.e. optimized and extended) on the slicing algorithm presented by Wang et al. [18]. They present a slicing method to reduce a single statechart based on dependence relations between states and transitions. We recall two of them [18].

Definition 1 (Parallel Data Dependence, $\rightarrow_{pdd}$). *A state u or transition r is (directly) parallel data dependent on a "concurrent" state v or transition t ($u \rightarrow_{pdd} v$ or $u \rightarrow_{pdd} t$, or $r \rightarrow_{pdd} v$, or $r \rightarrow_{pdd} t$) iff some output variables of the latter are used in the input of the other.*

Definition 2 (Synchronization Dependence, $\rightarrow_{sd}$). *A state u or transition r is (directly) synchronization dependent on a "concurrent" state v or transition t ($u \rightarrow_{sd} v$ or $u \rightarrow_{sd} t$, or $r \rightarrow_{sd} v$, or $r \rightarrow_{sd} t$) iff some events generated by the latter are used as trigger events of the other.*

Example 1. In the statechart of the washing machine (Fig. 3), $GetSoap \rightarrow_{pdd} t11$ due to the shared variable `content_soap`. $t11 \rightarrow_{sd} t2$ since the trigger event of $t11$ is generated by the action list of $t2$.

Suppose we want to verify the property *F(WashingCycle=GetSoap ∧ filling ≠ ∅)*: "eventually we reach the state GetSoap with a non-empty soap dosis.". Obviously, regions `Mode` and `Lamp` are irrelevant during the verification of this property; reaching the state `GetSoap` and acquiring `soap` is independent both of the program (mode) followed and of the fact the lamp is either blinking or not. Therefore, [18] slices these regions correctly away. Unfortunately, during the construction of its slice, [18] uses a false dependency: $GetSoap \rightarrow_{\text{pdd}} t11$. Consequently, region Soap will be added to the final slice. But this is a wrong way of working as there is no possible execution where the content_soap = full statement (i.e. action of $t11$) has an influence on filling = content_soap (i.e. action of $GetSoap$). Figure 5 motivates that the latter always executes before the first. SV$_t$L detects and removes such false dependencies; it avoids adding region Soap to the slice; it returns smaller slices than [18] and thus fastens verification even more. For a comparison between [18] and SV$_t$L we refer to [10].

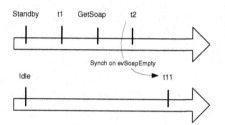

Fig. 5. Execution Timeline

Getting Rid of False Dependencies like $GetSoap \rightarrow_{\text{pdd}} t11$. Basically, SV$_t$L uses the following rule while searching for false parallel data dependencies:

$$x \nrightarrow_{\text{pdd}} y \text{ iff } x \rightarrow_{\text{so}} y$$

where $\rightarrow_{\text{so}}$ is a Lamport-like [9] happens-before relation on statecharts i.e. SV$_t$L ensures to take the execution chronology into account during the detection of a parallel data dependency. For our example SV$_t$L finds $GetSoap \nrightarrow_{\text{pdd}} t11$ iff it can derive that $GetSoap \rightarrow_{\text{so}} t11$. The happens-before relation $\rightarrow_{\text{so}}$ is defined as a combination of two other relations: the statechart concurrent order relation (Sect. 4.2) and the statechart sequential order (Sect. 4.3).

4.1 Preliminary Definitions

Let $F = \{WashingCycle, Mode, Lamp, Soap\}$ be the set of sequential automata the statechart is composed of. Let $A = (\sigma_A, s_A^0, \delta_A) \in F$ be a sequential automaton with σ_A the set of states of A; s_A^0 the initial state of A, and δ_A the set of transitions of A.

Definition 3 (Mutilated Automaton of A Ending in x). *Let $A \in F$ and $x \in (\sigma_A \cup \delta_A)$. The mutilated automaton of A ending in x, denoted as MAE_x, is the maximal (near)[3] sub-automaton of A ending in x (x included), i.e. x can be reached from each state and each transition of MAE_x.*

---

[3] Note that MAE_x and MAS_x are not complete automata if x belongs to δ_A.

Example 2. The mutilated automaton of `WashingCycle` (Fig. 3) ending in transition $t2$ (MAE_{t2}) is given by the set $\{Standby, t1, GetSoap, t2\}$.

Definition 4 (Predecessors of an Element ($pred_{element}$)). *Let $A \in F$ and $x \in (\sigma_A \cup \delta_A)$. Let MAE_x be the mutilated automaton of A ending in x. Let SCC_x ($\subseteq MAE_x$) be a strongly connected component of A containing x. Then, the set of all predecessors of element x is defined as follows:*

$$pred_x := \begin{cases} MAE_x \setminus \{x\} & \text{if } SCC_x = \emptyset \\ MAE_x \setminus SCC_x & \text{otherwise} \end{cases}$$

Example 3. For all possible executions of the orthogonal region `WashingCycle`, the predecessor set of `t2` is given by the set $pred_{t2} = \{Standby, t1, GetSoap\}$. The predecessor set of state `Off` (region `Lamp`) is empty as $MAE_{Off} = \{Off, On, t7, t8\} = SCC_{Off}$.

In a similar way, the successors of an element (state/transition) can be defined. We refer to [10] for further details.

4.2   Concurrent Order

SV_tL uses a concurrent order relation, denoted as $\xrightarrow{C}_{so}$, to represent the relative timing of concurrent states/transitions in a single statechart. Such a relation immediately follows from the internal broadcasting (synchronization) mechanism used in statecharts (see also Fig. 5); i.e. SV_tL directly derives an order relation between those concurrent states/transitions that communicate through events (messages), simply because "a message cannot be delivered before its sending". In [10] we have defined several concurrent order relations, but for the given example we only need to define the order relation between two concurrent transitions formally:

Definition 5. *The relation $\xrightarrow{C}_{so}$ on states and transitions of concurrent automata is a relation satisfying the following condition: if $A, B \in F$ are concurrent regions, $r \in \delta_A$, $t \in \delta_B$, and if $r \to_{sd} t$ (Definition 2) then $t \xrightarrow{C}_{so} r$, for all possible parallel executions of A and B.*

Example 4. $t2 \xrightarrow{C}_{so} t11$ since $t11 \to_{sd} t2$ (Fig. 3). As long as the trigger event of $t11$ has not been generated by $t2$, $t11$ never has the chance to be fired.

4.3   Sequential Order

SV_tL only manages to derive that $GetSoap \to_{so} t11$ if it also can rely upon a sequential order relation (see also Fig. 5). The sequential order, denoted as $\xrightarrow{S}_{so}$, is formally defined as follows:

Definition 6. *The relation $\xrightarrow{S}_{so}$ on states and transitions of an automaton is a relation satisfying the following condition: if $A \in F, x \in (\sigma_A \cup \delta_A)$ then (1)*

$pred_x \xrightarrow{S}_{so} x^4$, (2) $x \xrightarrow{S}_{so} succ_x{}^5$, for all possible executions of A i.e. no predecessor (successor) ever has the chance to be a successor (predecessor) as well,
no matter how A transitions between states.

Example 5. All the elements of $pred_{t2} = \{Standby, t1, GetSoap\}$ are guaranteed
to execute before t2 (Fig. 3).

4.4 Happens-Before on Statecharts

We define the relation $\rightarrow_{so} = (\xrightarrow{S}_{so} \cup \xrightarrow{C}_{so})^+$ to be our happens-before relation
on statecharts. With such a relation SV$_t$L establishes a useful ordering among
states and transitions of the statechart. The transitivity [10] of this order relation
imposes a trivial but interesting property due to definitions of predecessors and
successors.

Property 1. Let $A, B \in F$ be concurrent regions, $x \in (\sigma_A \cup \delta_A), y \in (\sigma_B \cup \delta_B)$.
The relation $\rightarrow_{so}$ on states and transitions of concurrent automata is a binary
relation with the following property: if $x \rightarrow_{so} y$ then (1) $pred_x \rightarrow_{so} y$, (2)
$x \rightarrow_{so} succ_y$, and (3) $pred_x \rightarrow_{so} succ_y$.

SV$_t$L now applies this property on Example 4 which results in $pred_{t2} \rightarrow_{so} t11$
and thus $\{Standby, t1, GetSoap\} \rightarrow_{so} t11$. It has derived that $GetSoap \rightarrow_{so} t11$,
so it safely concludes that $GetSoap$ never can be parallel data dependent on
$t11$. The slicer of SV$_t$L reduces the number of parallel dependencies used in slicing statecharts with concurrent states, and consequently avoids that irrelevant
information (e.g. region Soap) is added to the final slice.

5 Slicing the Multi-threaded Behavioral Model

As far as we know, slicing a multi-threaded model like ours never has been considered in the literature. The most straightforward way to slice such models is to
reduce the model to a single statechart. We have some arguments to avoid this.

- In a multi-threaded design model components are allowed to either communicate asynchronously with signal events, or synchronously with call events, or
  even both. If statecharts communicate synchronously, the sender is blocked
  until the receiver has returned an answer. The UML semantics forbids that
  the elements (states, transitions) of a single statechart communicate through
  synchronous calls because both of the re-entrance problem as well as the
  blocking mechanism.
- To build a single statechart for the multi-threaded design model, each thread
  will correspond to a separate concurrent region. But how does the behavior
  of these regions look like as a thread groups together a collection of reactive
  objects (i.e. each object has its behavior expressed in a different statechart)?

---

[4] $\forall y \in pred_x : y \xrightarrow{S}_{so} x$

[5] $\forall y \in succ_x : x \xrightarrow{S}_{so} y$

- And if we manage to build a single statechart, then there is still the fact the slicer will add to much irrelevant information to the final slice. This follows from some dependence relations (which we don't have mentioned here) and some steps of the slicing algorithm (see [10]).
- Finally, we lose the object-oriented structure of the model. Side-effect: the transparency of the verification gets lost.

The above arguments strongly prevent us from generating a single statechart diagram. Our goal is now to define a *parallel slicing algorithm* on the given multi-threaded model so that in the final slice, irrelevant information comes along as little as possible.

5.1   Global Dependence Relations

To slice a collection of statechart diagrams, SV_tL must address the possible inter-actions between the statecharts, as they imply new dependence relations. Statecharts either communicate through *global variables* or *global events*. A global variable does not belong to any statechart in particular and can be accessed from any other statechart. Any statechart can modify global variables and any part (state/transition) of a statechart may depend on it. This results in a *global data dependence* (Definition 7). Obviously, something similar can be said of global events resulting in a *global synchronization dependence* (Definition 8). However, the synchronous call, as a global event, also implies a global data dependence due to the return value that is needed to de-block the sender of the call. Note that this corresponds to the locking conditions as presented in [8].

Definition 7 (Global Data Dependence, $\rightarrow_{gdd}$). *Let S_1, S_2 be two different statecharts of the input model. A state $u \in S_1$ or transition $r \in S_2$ is (directly) global data dependent on a state $v \in S_2$ or transition $t \in S_2$ ($u \rightarrow_{gdd} v$ or $u \rightarrow_{gdd} t$, or $r \rightarrow_{gdd} v$, or $r \rightarrow_{gdd} t$) iff*

- *some global output variables of the latter are used in the input of the other*
- *or some call events (triggers) of the latter are sent out by the other.*

Definition 8 (Global Synchronization Dependence, $\rightarrow_{gsd}$). *Let S_1, S_2 be two different statecharts of the input model. A state $u \in S_1$ or transition $r \in S_1$ is (directly) global synchronization dependent on a state $v \in S_2$ or transition $t \in S_2$ ($u \rightarrow_{gsd} v$ or $u \rightarrow_{gsd} t$, or $r \rightarrow_{gsd} v$, or $r \rightarrow_{gsd} t$) iff some global events generated by the latter are used as trigger events of the other.*

Example 6. $t12 \rightarrow_{gsd} GetWater$ (Fig. 3- 4) since the trigger event of $t12$ is global event that is generated at the moment $GetWater$ is entered.

Note that the multi-threaded behavioral model has other global dependence relations but we omit to mention them here.

5.2 A Directed Graph

SV$_t$L does not use the global dependence relations to slice the set of statecharts in parallel. Instead, it uses these global relations to connect the statecharts to each other in such a way that it obtains a graph-like structure. It is sufficient to draw a global directed edge for each global dependence relation. With these edges we devise an elegant parallel slicing algorithm (Section 5.3). The edges are formally defined as follows:

Definition 9 (Global Data Edge, $dEdge(source, target)$). *Let S_1, S_2 be two different statecharts of the input model. There exists a global data edge from $x \in S_1$ to $y \in S_2$ ($dEdge(x,y)$) iff $x \rightarrow_{gdd} y$.*

Definition 10 (Global Synchronization Edge, $sEdge(source, target)$). *Let S_1, S_2 be two different statecharts of the input model. There exists a global synchronization edge from $x \in S_1$ to $y \in S_2$ ($sEdge(x,y)$) iff $x \rightarrow_{gsd} y$.*

Example 7. There exist a global synchronization edge between $t12$ and *Get − Water*, notation $sEdge(t12, GetWater)$, since $t12 \rightarrow_{gsd} GetWater$ (see Ex. 6).

5.3 The Parallel Algorithm

Once having constructed the graph, SV$_t$L is able to slice the collection of statecharts independently. This means that in the worst case, SV$_t$L has to execute three slicing algorithms in parallel: a slicer for the statechart of the washing machine (Fig. 3), a slicer for the statechart of the water tank (Fig. 4), and finally a slicer for the statechart of Marvin.

Suppose we want to verify the property *F (Watertank = Filling): "eventually the water tank is filled with water"*. Such a property starts a first slicing algorithm (*Water Tank Slicer*) to possibly reduce the statechart of the water tank. SV$_t$L adds transition $t12$ to the slice, since we have to be able to reach state *Filling*, which is of course also part of the final slice. At this point, SV$_t$L initializes a second slicing algorithm (*Washing Machine Slicer*), as there exists a global synchronization edge between $t12$ and *GetWater* (see Ex. 7). Thus, global edges cause the adding information into the slice of other statecharts. Both slicers continue working in parallel. The water tank slicer is not able to remove some elements of the statechart, while the washing machine slicer reduces the statechart to a sequential statechart consisting of the hierarchical state `WashingCycle`. Note that SV$_t$L didn't allow to let Marvin play a role here, so its statechart will not belong to the final slice.

Remark. The happens-before relation on a single statechart can easily be lifted to a happens-before relation on a set of collaborating statecharts. This reduces the amount of global dependence relations (and consequently the number of global directed edges) so that SV$_t$L retrieves smaller slices.

6   The Verification Model: Underlying Semantics

Once SV_tL has reduced the input model, it translates this sliced model into the input language of the model checker. Thereafter, SV_tL instructs the model checker to verify the property. A necessary first step towards the construction of the verification model is the definition of a formal semantics of the behavioral model yielding finite Kripke structures. This issue is addressed in many research papers [1, 2, 15, 11] , but the suggested approaches are unsatisfying with respect to multi-threaded applications; they cover only statechart diagrams in isolation. Intuitively, the Kripke structure for a multi-threaded behavioral model is a combination of the Kripke structures that we can build for each statechart separately.

Example 8. The Kripke structure for the model given in Fig. 2 is a combination of the Kripke structures belonging to the `WashingMachine`, the `Heater`, the `Tank`, and `Marvin` respectively.

UML defines the semantics of a statechart as a run-to-completion step (RTC-step) semantics. Such a semantics states that an event can only be dispatched when the processing of the previous event has been completed. When targeting multi-threaded applications, we have to lift the semantics to a thread-run-to-completion step semantics, while still allowing that each object belonging to the thread performs a run-to-completion algorithm. Due to the presence of several objects, an event is either dispatched from the call queue (q_c, synchronous calls) or the signal queue (q_e, asynchronous signals) and sent to the statechart of the object that responds to the message; which is of course unique. Thereafter, the target statechart is repeatedly evaluated until a stable configuration (no triggerless (ε-) transitions can be taken) is reached, i.e. it executes an RTC-step. Only then, the event is fully consumed and the next event can be dispatched, and possibly another statechart responds to the latter event. Formally:

Definition 11 $(\overset{STEP}{\rightarrow})$. *Let 'last' be the statechart to which an event is last dispatched. Let 'auto' be one of the objects that became unstable (triggerless (ε-) transitions can be taken) due to the progress made in 'last'. Then, the transition relation of a thread is defined as follows:*

$$\overset{STEP}{\rightarrow} = \begin{cases} 1. & (\overset{prog}{\longrightarrow}_\varepsilon)last & if\ (MaxET(\mathcal{C})_{dispatch(\mathcal{C})=\varepsilon})last \neq \emptyset \\ 2. & (\overset{prog}{\longrightarrow}_\varepsilon)auto & if\ (MaxET(\mathcal{C})_{dispatch(\mathcal{C})=\varepsilon})auto \neq \emptyset \\ 3. & \overset{prog}{\longrightarrow}_{q_c[0]} & if\ MaxET(\mathcal{C})_{dispatch(\mathcal{C})=q_c[0]} \neq \emptyset \\ 4. & \overset{prog}{\longrightarrow}_{q_e[0]} & if\ MaxET(\mathcal{C})_{dispatch(\mathcal{C})=q_e[0]} \neq \emptyset \\ 5. & \overset{stut}{\longrightarrow} & otherwise \end{cases}$$

with $MaxET(\mathcal{C})$ the set of maximal enabled transitions
with $\overset{prog}{\longrightarrow}_\varepsilon$: progress due to a triggerless transition
with $\overset{prog}{\longrightarrow}_{q_c[0]}$: progress due to a synchronous call

with $\xrightarrow{prog}_{q_e[0]}$: *progress due to an asynchronous signal*

with $\xrightarrow{stut}$: *lack of progress or exception, guarantees a total transition relation.*

However, before serving a new event, the dispatcher needs to verify whether there are other objects that became unstable due to the progress made in the last evaluated statechart.

7   What About Model Equivalences?

Original Statechart Model ≡ Sliced Statechart Model. Slicing an EHA H (= the behavioral model) with respect to a LTL (CTL) property φ should yield a smaller residual EHA H_s that preserves and reflects the satisfaction of φ and has as little irrelevant information as possible. In case of a LTL$_X$[6] specification, H and H_s have to be φ-*stuttering equivalent* [10]; while in case of a CTL$_X$ specification, the models have to be φ-*stuttering bisimular* [10].

Sliced Statechart Model ≡ Kripke Model. Intuitively, a statechart model is equivalent to a Kripke model if they have the same semantics i.e. if they denote the same behavior. To verify the transformation, a semantics for both the statechart model and the Kripke model with the same semantic domain is required. An elegant way is to define the semantics as coalgebras [13, 6]. Consequently, the proof of the behavioral equivalence is reduced to the construction of a *coalgebraic bisimulation relation* between the statechart model and the Kripke model.

8   Conclusion

The more complex systems of today require modeling methods and tools that allow errors to be detected in the initial phases of development. This paper has presented a prototype tool, SV$_t$L, that enables the behavior of the system expressed in UML to be verified in a completely automatic way based on model checking and slicing. This paper has also shortly discussed the correctness of the verification approach covered by SV$_t$L.

SV$_t$L carries out a formal logical framework in which to verify UML statechart diagrams. The advantage of such an approach is that at the one hand software developers can use UML to specify and to describe the software, and on the other hand, they benefit from a "well-known" efficient verification method.

As there exist already similar UML verification tools that act as usable frontends to existing model checkers, SV$_t$L is the first tool that integrates a slicing step while verifying UML statechart models. Its strength, the slicing-based verification environment, makes the verification approach applicable to multi-threaded UML statechart models. We believe that SV$_t$L is valuable enough to

---

[6] LTL$_X$ (CTL$_X$): LTL (CTL) without the next **X** operator.

encourage the use of formal methods by software engineers. SV_tL makes the software development process more effective, as it can be repeatedly activated to verify behavioral designs.

Currently we are working on an evaluation over one or more "real-life" case studies to prove the usefulness of the slicing algorithm.

References

[1] Michelle L. Crane and Juergen Dingel. On the Semantics of UML State Machines: Categorization and Comparison. Technical Report 2005, School of Computing Queen's University Kingston, Ontario, Canada, 2005-501.

[2] Alexandre David, Johann Deneux, and Julien d'Orso. A Formal Semantics for UML Statecharts. Technical Report 2003-010, Uppsala University, 2003.

[3] Matthew B. Dwyer, George S. Avrunin, and James C. Corbett. Patterns in Property Specifications for Finite-State Verification. In *ICSE '99: Proceedings of the 21st international conference on Software engineering*, pages 411–420, Los Alamitos, CA, USA, 1999. IEEE Computer Society Press.

[4] Stefania Gnesi. Model Checking of Embedded Systems, Januari 2003.

[5] Maria Encarnación Beato Gutiérrez, Manuel Barrio-Solórzano, Carlos Enrique Cuesta Quintero, and Pablo de la Fuente. UML Automatic Verification Tool with Formal Methods. *Electr. Notes Theor. Comput. Sci.*, 127(4):3–16, 2005.

[6] Bart Jacobs. Many-Sorted Coalgebraic Modal Logic: a Model-Theoretic Study. *ITA*, 35(1):31–59, 2001.

[7] Steven D. Johnson. Formal Methods in Embedded Design. *Computer*, 36(11):104–106, 2003.

[8] Vineet Kahlon and Aarti Gupta. An Automata-Theoretic Approach for Model Checking Threads for LTL properties. In *LICS '06: Proceedings of the 21st Annual IEEE Symposium on Logic in Computer Science*, pages 101–110, Washington, DC, USA, 2006. IEEE Computer Society.

[9] Leslie Lamport. Time, Clocks, and the Ordering of Events in a Distributed System. *Commun. ACM*, 21(7):558–565, 1978.

[10] Sara Van Langenhove. Towards the Correctness of Software Behavior in UML: A Model Checking Approach Based on Slicing. Ph.D. Thesis UGent, May 2006.

[11] Diego Latella, Istvan Majzik, and Mieke Massink. Towards a Formal Operational Semantics of UML Statechart Diagrams. In *Proc. FMOODS'99, IFIP TC6/WG6.1 Third International Conference on Formal Methods for Open Object-Based Distributed Systems, Florence, Italy, February 15-18, 1999*, pages 331–347. Kluwer, B.V., 1999.

[12] Kenneth L. McMillan. Cadence SMV. Available from http://embedded.eecs.berkeley.edu/Alumni/kenmcmil/smv/.

[13] Sun Meng, Zhang Naixiao, and Luis S. Barbosa. On Semantics and Refinement of UML Statecharts: A Coalgebraic View. *SEFM '04: Proceedings of the Software Engineering and Formal Methods, Second International Conference on (SEFM'04)*, 00:164–173, 2004.

[14] Erich Mikk, Yassine Lakhnech, Michael Siegel, and Gerard J. Holzmann. Implementing Statecharts in Promela/SPIN. In *Proceedings of the 2nd IEEE Workshop on Industrial-Strength Formal Specification Techniques*, pages 90–101. IEEE Computer Society, October 1998.

[15] Ivan Paltor and Johan Lilius. Formalising UML State Machines for Model Checking. In *UML*, pages 430–445, 1999.

[16] Ivan Porres Paltor and Johan Lilius. vUML: A Tool for Verifying UML Models. In Robert J. Hall and Ernst Tyugu, editors, *Proc. of the 14th IEEE International Conference on Automated Software Engineering, ASE'99*. IEEE, 1999.

[17] G. Ramalingam. Context-Sensitive Synchronization-Sensitive Analysis is Undecidable. *ACM Trans. Program. Lang. Syst.*, 22(2):416–430, 2000.

[18] Ji Wang, Wei Dong, and Zhi-Chang Qi. Slicing Hierarchical Automata for Model Checking UML Statecharts. In *ICFEM '02: Proceedings of the 4th International Conference on Formal Engineering Methods*, volume 2495 of *Lecture Notes in Computer Science*, pages 435–446, London, UK, 2002. Springer.

A (Co)Algebraic Analysis of Synchronization in CSP⋆

Uwe Wolter

Department of Informatics, University of Bergen, 5020 Bergen, Norway
`wolter@ii.uib.no`

Abstract. We present a model theoretic analysis of synchronization of deterministic CSP processes. We show that there is co-amalgamation within the indexed coalgebraic reconstruction of CSP developed in [14]. Synchronization, however, can not be characterized in terms of co-amalgamation. We show that synchronization can be described, nevertheless, as a pullback construction within the corresponding fibred algebraic setting. Analyzing and generalizing the transition between the indexed and the fibred setting we show that for a wide range of signature embeddings $\varphi : \Sigma_1 \to \Sigma_2$ the Σ_1-algebras, traditionally considered as parameter algebras, can be considered also as signatures, instead.

1 Introduction

We present a further outcome of a more comprehensive program of "Dualizing Universal Algebra" [7,11,13,14,15]. Having in mind that structuring and modularization in Algebraic Specifications is based on amalgamation [3] we have been looking this time for a dualization of amalgamation. And, since synchronization is one of the most important structuring mechanisms in system specification and process calculi, we have also investigated the relation between co-amalgamation and synchronization.

An investigation of co-amalgamation has to be based on a coalgebraic exposition of system specifications. Therefore we chose for a first analysis the coalgebraic reconstruction of (deterministic) CSP [6] presented in [14].

For deterministic CSP, as for many other approaches to system specifications, sets of action/input symbols can be considered as signatures. And we can define coreduct functors and co-amalgamation within the corresponding indexed coalgebraic setting developed in [14]. Co-amalgamation, however turns out to be a very poor mechanism (in the same way as amalgamation is poor in case of one-sorted algebras) and, moreover, synchronization can not be characterized in terms of co-amalgamation.

Analysing the situation more thouroughly from an algebraic viewpoint we made two observations worth communicating:

- The possibility to take sets of input symbols as signatures is based on the fact that for a wide range of signature embeddings $\varphi : \Sigma_1 \to \Sigma_2$ the corresponding forgetful functor $U_\varphi : \mathbf{PAlg}(\Sigma_2) \to \mathbf{PAlg}(\Sigma_1)$ is a fibration.

⋆ Research partially supported by the Norwegian NFR project MoSIS/IKT.

J.L. Fiadeiro and P.-Y. Schobbens (Eds.): WADT 2006, LNCS 4409, pp. 156–170, 2007.

– Synchronization in CSP actually combines automata with idle actions and can be described by a pullback construction within the fibred algebraic setting that is obtained from the indexed coalgebraic setting by the *Grothendieck construction*. In the pullback construction, however, the idle actions don't play the same rôle as they do it, for example, in CommUnity [4]. That is, the pullback is not taken in the category of automata with idle actions.

The paper presents an intermediate state of our research. We hope, however, that it encourages also other computer scientists to analyse and to resolve the conceptual mismatch pointed out in the paper.

2   Deterministic Partial Automata

For a first analysis and since there is no clean model theoretic interpretation of non-deterministic CSP available (compare [14]) we restrict here to deterministic CSP. The insight that has been elaborated in [14] is that deterministic CSP is concerned with *deterministic partial automata without output*, i.e., with triples $\mathcal{M} = (I, S, d)$ where I is a set of *input symbols*, S a set of *states*, and $d : S \times I \nrightarrow S$ a partial *state transition function*. Note that $\mathcal{M}$ can be considered as a partial algebra with two sorts and one partial operation (compare [2,9,12]). It is well-known that for any such partial function there is an equivalent *curried* version, i.e., a total function $\lambda(d) : S \to [I \nrightarrow S]$ with $i \in \mathsf{dom}(\lambda(d)(s))$ iff $(s, i) \in \mathsf{dom}(d)$ for all $s \in S$, $i \in I$, and with $\lambda(d)(s)(i) = d(s, i)$ for all $i \in \mathsf{dom}(\lambda(d)(s))$. In such a way an automaton $\mathcal{M}$ can be described equivalently using the curried version of d by the triple $(I, S, \lambda(d))$.

The crucial point now is that those triples can be interpreted as $I_{\to}$-coalgebras [7,11] $(S, \lambda(d))$ for the functor $I_{\to} : \mathbf{Set} \to \mathbf{Set}$ where $I_{\to}(S) \overset{def}{=} [I \nrightarrow S]$ is the set of all partial functions from I into S and for any mapping $f : S_1 \to S_2$ the mapping $I_{\to}(f) : [I \nrightarrow S_1] \longrightarrow [I \nrightarrow S_2]$ is defined by post-composition

$$I_{\to}(f)(g) \overset{def}{=} g\,; f \quad \text{for all partial functions } g \in [I \nrightarrow S_1].$$

An $I_{\to}$-homomorphism $f : (S_1, \beta_1) \to (S_2, \beta_2)$ between two $I_{\to}$-coalgebras (S_1, β_1) and (S_2, β_2) is a mapping $f : S_1 \to S_2$ such that the following diagram commutes

$$
\begin{array}{ccc}
S_1 & \overset{\beta_1}{\longrightarrow} & [I \nrightarrow S_1] \\
{\scriptstyle f}\downarrow & & \downarrow{\scriptstyle -\,;\,f} \\
S_2 & \overset{\beta_2}{\longrightarrow} & [I \nrightarrow S_2]
\end{array}
$$

Note that for any g in $[I \nrightarrow S_1]$ $\mathsf{dom}I_{\to}(g) = \mathsf{dom}(g\,; f) = \mathsf{dom}(g)$ since $f : S_1 \to S_2$ is a total mapping. That is, $I_{\to}$-homomorphisms preserve and reflect definedness and correspond, in such a way, exactly to the closed homomorphisms [2,9,12] between the partial algebras given by the uncurried version of (S_1, β_1) and (S_2, β_2), respectively. The category of all $I_\rightarrow$-coalgebras and all $I_\rightarrow$-homomorphisms will be denoted by $\mathbf{DA}_I$.

A basic result in [14] is that the deterministic CSP processes can be characterized in terms of homomorphisms between coalgebras: A deterministic process with alphabet I is defined to be any *non-empty prefix closed* subset P of I^*, i.e., any subset $P \in I^*$ which satisfies the two conditions (i) $\langle\rangle \in P$, and (ii) $(\forall s, t \in I^* : s\hat{\ }t \in P \Rightarrow s \in P)$, where $\langle\rangle \in P$ denotes the empty *trace* (finite sequence) and $s\hat{\ }t$ the *catenation* of traces. We denote by DP_I the set of all deterministic processes with alphabet I and define a $I_\rightarrow$-coalgebra $\mathcal{H}_I = (DP_I, n_I)$ with $n_I : DP_I \rightarrow [I \nrightarrow DP_I]$ where for any $P \in DP_I$ the domain of $n_I(P)$ is denoted in [6] by P^0 and defined by $\mathsf{dom}(n_I(P)) = \{a \mid \langle a \rangle \in P\}$. $n_I(P)(a)$ for any $a \in P^0 = \mathsf{dom}(n_I(P))$ is denoted in [6] by $P(a)$ and defined by $n_I(P)(a) = \{t \mid \langle a \rangle\hat{\ }t \in P\}$.

The *Hoare-model* $\mathcal{H}_I$ is characterized uniquely, up to isomorphism, by the property of being the final object in $\mathbf{DA}_I$: Let $\mathcal{M} = (S, \beta)$ be an $I_\rightarrow$-coalgebra. We write $s \xrightarrow{a} s'$ for $a \in \mathsf{dom}(\beta(s))$ and $\beta(s)(a) = s'$. The process that can be observed in $\mathcal{M}$ *starting in a state* s is given by

$$\tau_\mathcal{M}(s) \overset{def}{=} \{\langle\rangle\} \cup \{\langle a_1, \ldots, a_n \rangle \mid s \xrightarrow{a_1} s_1 \xrightarrow{a_2} \cdots \xrightarrow{a_n} s_n\} \in DP_I$$

In analogy to [10] $\tau_\mathcal{M}(s)$ could be also called the *language accepted by* s.

It is easy to see that the mapping $\tau_\mathcal{M} : S \rightarrow DP_I$ defines an $I_\rightarrow$-homomorphisms $\tau_\mathcal{M} : \mathcal{M} \rightarrow \mathcal{H}_I$ and the uniqueness of $\tau_\mathcal{M}$ can be shown straightforwardly by induction on the length of traces (see [14]).

3   Concurrent Interaction and Synchronization

The concurrent interaction $P \parallel Q$ of processes P and Q with different alphabets I and J reflects, according to [6], *lock-step synchronization*. Only events that are in both alphabets, i.e., in the intersection $I \cap J$, are required to synchronize. However, events in the alphabet of P but not in the alphabet of Q may occur independently of Q whenever P engages in them. Similarly, Q may engage alone in events which are in the alphabet of Q but not of P. In such a way the alphabet of the process $P \parallel Q$ will be the union $I \cup J$ of the alphabets of the component processes. To model this kind of synchronization we have to define a map $\_ \parallel \_ : DP_I \times DP_J \rightarrow DP_{I \cup J}$.

In Section 2 we have seen that the Hoare-model $\mathcal{H}_K = (DP_K, n_K)$ is a final $K_\rightarrow$-coalgebra. This allows us to define mappings from an arbitrary set S into DP_K *coinductively* [7,11]: We have only to construct a $K_\rightarrow$-coalgebra $\mathcal{M} = (S, \beta)$ with carrier S. Then, by finality of $\mathcal{H}_K$, there exists a unique $K_\rightarrow$-homomorphism $\tau_\mathcal{M} : \mathcal{M} \rightarrow \mathcal{H}_K$. The substantial problem is to design $\mathcal{M}$ in such a way that the underlying mapping $\tau_\mathcal{M} : S \rightarrow DP_K$ becomes the intended one.

The coinductive definition of the intended mapping $\_ \parallel \_ : DP_I \times DP_J \rightarrow DP_{I \cup J}$ can be extracted from law 7, page 71 in [6]. The synchronization of $\mathcal{H}_I$ and $\mathcal{H}_J$ provides an $(I \cup J)_\rightarrow$-coalgebra

$$\mathcal{S}_{I,J} = (DP_I \times DP_J, syn_{I,J} : DP_I \times DP_J \longrightarrow [I \cup J \twoheadrightarrow DP_I \times DP_J])$$

as follows: For any pair of processes $(P, Q) \in DP_I \times DP_J$ we define

$$\mathsf{dom}(syn_{I,J}(P, Q))$$
$$\overset{def}{=} (\mathsf{dom}(n_I(P)) \setminus J) \cup (\mathsf{dom}(n_I(P)) \cap \mathsf{dom}(n_J(Q))) \cup (\mathsf{dom}(n_J(Q)) \setminus I)$$

and for any $c \in \mathsf{dom}(syn_{I,J}(P, Q))$ we set

$$syn_{I,J}(P, Q)(c) \overset{def}{=} \begin{cases} (n_I(P)(c), Q) & , c \in \mathsf{dom}(n_I(P)) \setminus J \\ (n_I(P)(c), n_J(Q)(c)), & c \in \mathsf{dom}(n_I(P)) \cap \mathsf{dom}(n_J(Q)) \\ (P, n_I(Q)(c)) & , c \in \mathsf{dom}(n_J(Q)) \setminus I \end{cases}$$

The final $(I \cup J)_{\twoheadrightarrow}$-homomorphism $\tau_{\mathcal{S}_{I,J}} : \mathcal{S}_{I,J} \to \mathcal{H}_{I \cup J}$ provides the intended concurrent interaction operator $\_ \parallel \_ : DP_I \times DP_J \to DP_{I \cup J}$.

This coalgebraic reconstruction of synchronization in CSP makes it possible to extend synchronization to arbitrary automata:

Definition 1. *For partial automata* $\mathcal{M}_1 = (S_1, \beta_1 : S_1 \to [I \twoheadrightarrow S_1])$ *and* $\mathcal{M}_2 = (S_2, \beta_2 : S_2 \to [J \twoheadrightarrow S_2])$ *we define the corresponding* synchronized automaton

$$\mathcal{M}_1 \parallel \mathcal{M}_2 = (S_1 \times S_2, syn_{\mathcal{M}_1, \mathcal{M}_2} : S_1 \times S_2 \longrightarrow [I \cup J \twoheadrightarrow S_1 \times S_2])$$

as follows: For each $(s_1, s_2) \in S_1 \times S_2$ *we define*

$$\mathsf{dom}(syn_{\mathcal{M}_1, \mathcal{M}_2}(s_1, s_2))$$
$$\overset{def}{=} (\mathsf{dom}(\beta_1(s_1)) \setminus J) \cup (\mathsf{dom}(\beta_1(s_1)) \cap \mathsf{dom}(\beta_2(s_2))) \cup (\mathsf{dom}(\beta_2(s_2)) \setminus I)$$

and for any $c \in \mathsf{dom}(syn_{\mathcal{M}_1, \mathcal{M}_2}(s_1, s_2))$ *we set*

$$syn_{\mathcal{M}_1, \mathcal{M}_2}(s_1, s_2)(c) \overset{def}{=} \begin{cases} (\beta_1(s_1)(c), s_2) & , c \in \mathsf{dom}(\beta_1(s_1)) \setminus J \\ (\beta_1(s_1)(c), \beta_2(s_2)(c)), & c \in \mathsf{dom}(\beta_1(s_1)) \cap \mathsf{dom}(\beta_2(s_2)) \\ (s_1, \beta_2(s_2)(c)) & , c \in \mathsf{dom}(\beta_2(s_2)) \setminus I \end{cases}$$

Now, it turns out that synchronization is indeed the semantical basis of concurrent interaction. That is, concurrent interaction of processes describes exactly how the processes in an arbitrary synchronized automaton $\mathcal{M}_1 \parallel \mathcal{M}_2$ can be reconstructed from the processes of the single automata $\mathcal{M}_1$ and $\mathcal{M}_2$. In other words, synchronization of automata can be seen as a compatible semantical extension of concurrent interaction of processes as stated in

Theorem 1. *For any partial automata* $\mathcal{M}_1 = (S_1, \beta_1 : S_1 \to [I \twoheadrightarrow S_1])$, $\mathcal{M}_2 = (S_2, \beta_1 : S_2 \to [J \twoheadrightarrow S_2])$, *and any pair of states* $(s_1, s_2) \in S_1 \times S_2$ *we have that*

$$\tau_{\mathcal{M}_1 \parallel \mathcal{M}_2}(s_1, s_2) = \tau_{\mathcal{S}_{I,J}}(\tau_{\mathcal{M}_1}(s_1), \tau_{\mathcal{M}_2}(s_2)) = \tau_{\mathcal{M}_1}(s_1) \parallel \tau_{\mathcal{M}_2}(s_2)$$

Proof: We have to show that the following diagram commutes.

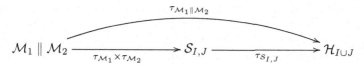

Since $_-\| \,_- : DP_I \times DP_J \to DP_{I \cup J}$ is given by the final $(I \cup J)_{\to}$-homomorphism $\tau_{\mathcal{S}_{I,J}} : \mathcal{S}_{I,J} \to \mathcal{H}_{I \cup J}$ it suffices to show that the mapping $\tau_{\mathcal{M}_1} \times \tau_{\mathcal{M}_2} : S_1 \times S_2 \to DP_I \times DP_J$ constitutes a $(I \cup J)_{\to}$-homomorphism $\tau_{\mathcal{M}_1} \times \tau_{\mathcal{M}_2} : \mathcal{M}_1 \| \mathcal{M}_2 \to \mathcal{S}_{I,J}$ (see [14] for a proof). Then the diagram commutes due to the uniqueness of $(I \cup J)_{\to}$-homomorphisms into $\mathcal{H}_{I \cup J}$. $\qquad\qquad\square$

4   Co-amalgamation

In universal coalgebra, *abstract signatures* are functors from **Set** into **Set**; thus, *abstract signature morphims* are given by natural transformations [11]. That is, in coalgebraic specification formalisms the corresponding category **Sig** of signatures is usually assumed to be a subcategory of the functor category **Func(Set, Set)**. And the duality between coalgebras and algebras is reflected by the fact that we have a covariant model functor $mod : \mathbf{Sig} \to \mathbf{Cat}$ instead of a contravariant model functor $mod : \mathbf{Sig}^{op} \to \mathbf{Cat}$ as in algebraic specifications.

The interesting point, in case of CSP, is that abstract signatures $I_{\to} : \mathbf{Set} \to \mathbf{Set}$ are represented by sets I and that maps $\phi : I \to J$ represent the abstract signature morphisms, i.e., natural transformations $\phi_{\to} : J_{\to} \Rightarrow I_{\to}$ in the opposite direction, where the components

$$\phi_{\to}(S) \stackrel{def}{=} (\phi; \_) : [J \twoheadrightarrow S] \to [I \twoheadrightarrow S]$$

are simply given by pre-composition. In other words: The assignments $I \mapsto I_{\to}$ and $\phi \mapsto \phi_{\to}$ define an embedding of $\mathbf{Set}^{op}$ into **Func(Set, Set)**. To have a basis for dualizing amalgamation, however, we take as our category **Sig** of abstract signatures instead of **Set** (or $\mathbf{Set}^{op}$) the subcategory of **Func(Set, Set)** given by the image of this embedding.

Any abstract signature morphism $\phi_{\to} : J_{\to} \Rightarrow I_{\to}$, i.e., any map $\phi : I \to J$, defines now a *coreduct functor* $\mathbf{DA}_{\phi} : \mathbf{DA}_J \to \mathbf{DA}_I$ with

$$\mathbf{DA}_{\phi}(A, \alpha) \stackrel{def}{=} (A, \phi; \alpha(\_)) \quad \text{for any } J_{\to}\text{-coalgebra } (A, \alpha).$$

and $\mathbf{DA}_{\phi}(f) \stackrel{def}{=} f$ for any $J_{\to}$-homomorphism $f : (A, \alpha) \to (B, \beta)$

$$
\begin{array}{ccccc}
A & \xrightarrow{\;\alpha\;} & [J \twoheadrightarrow A] & \xrightarrow{\;\phi;\_\;} & [I \twoheadrightarrow A] \\
{\scriptstyle f}\downarrow & & \downarrow{\scriptstyle \_;f} & & \downarrow{\scriptstyle \_;f} \\
B & \xrightarrow{\;\beta\;} & [J \twoheadrightarrow B] & \xrightarrow{\;\phi;\_\;} & [I \twoheadrightarrow B]
\end{array}
$$

That is, we delete all transitions in the partial automaton (A, α) labelled by elements in $J \backslash \phi(I)$, and the other transitions are multiplied, if ϕ is non-injective, and the labels are renamed according to ϕ.

It is straightforward to show that the assigments $I_{\to} \mapsto \mathbf{DA}_I$ and $\phi_{\to} \mapsto \mathbf{DA}_{\phi}$ define a covariant model functor $\mathbf{DA} : \mathbf{Sig} \to \mathbf{Cat}$

In algebraic specifications we have amalgamation if the corresponding contravariant model functor maps pushouts in **Sig** into pullbacks in **Cat**.

Dually we have, in our case, co-amalgamation if **DA** maps pullbacks in **Sig** into pullbacks in **Cat** or, equivalently formulated, if a pushout diagram of maps between sets of input symbols induces a pullback diagram in **Cat**:

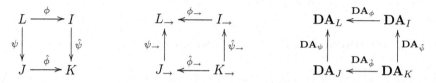

To prove that the diagram on the right is a pullback in **Cat** we have to show that for any $I_\rightarrow$-coalgebra (A, α) and any $J_\rightarrow$-coalgebra (B, β) with $\mathbf{DA}_\phi(A, \alpha) = \mathbf{DA}_\psi(B, \beta)$ there is a unique $K_\rightarrow$-coalgebra (C, γ) such that $\mathbf{DA}_{\hat{\psi}}(C, \gamma) = (A, \alpha)$ and $\mathbf{DA}_{\hat{\phi}}(C, \gamma) = (B, \beta)$. Due to the definition of the coforgetful functors, however, this means that $(C, \hat{\psi}; \gamma(\_)) = (A, \alpha)$ and $(C, \hat{\phi}; \gamma(\_)) = (B, \beta)$. That is, we require $A = B = C$ and this set has to be also the carrier of the $L_\rightarrow$-coalgebra $\mathbf{DA}_\phi(A, \alpha) = \mathbf{DA}_\psi(B, \beta)$.

In such a way, the assertion we have to show reduces to: For any maps $\alpha : C \rightarrow [I \twoheadrightarrow C]$ and $\beta : C \rightarrow [J \twoheadrightarrow C]$ with $\phi; \alpha(\_) = \psi; \beta(\_) : C \rightarrow [L \twoheadrightarrow C]$ there exists a unique $\gamma : C \rightarrow [K \twoheadrightarrow C]$ such that $\hat{\psi}; \gamma(\_) = \alpha$ and $\hat{\phi}; \gamma(\_) = \beta$. This assertion, however, is immediately ensured since the inner square in the diagram below is a pullback in **Set**.

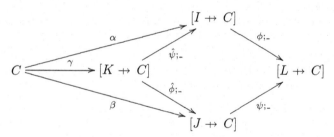

That is, the co-amalgamated automaton (C, γ) has the same carrier as the two given automata (C, α) and (C, β) and for any $c \in C$ the corresponding one-step transition $\gamma(c) : K \twoheadrightarrow C$ is given by a compatible union of the two one-step transitisions $\alpha(c) : I \twoheadrightarrow C$ and $\beta(c) : J \twoheadrightarrow C$ due to the pushout property of the inner square in the diagram below (bear in mind that the embedding of **Set** into the category **Par** of partial maps preserves pushouts).

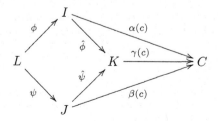

We is evident that co-amalgamation of automata is a very poor mechanism, however, in the same way as amalgamation is poor in one-sorted algebraic specifications where amalgamation just collects the different operations over the same carrier from two given algebras.

Based on the pullback and pushout property, respectively, also homomorphisms between coalgebras can be co-amalgamated.

5   Synchronization and Co-amalgamation

If we look now at synchronization in view of co-amalgamation we can make the simple observation that the diagram on the left below is a pushout diagram in **Set** thus we obtain a corresponding induced pullback diagram in **Cat**.

$$
\begin{array}{ccc}
I \cap J & \xrightarrow{\phi} & I \\
\psi \downarrow & & \downarrow \hat{\psi} \\
J & \xrightarrow{\hat{\phi}} & I \cup J
\end{array}
\qquad\qquad
\begin{array}{ccc}
\mathbf{DA}_{I \cap J} & \xleftarrow{\mathbf{DA}_{\phi}} & \mathbf{DA}_I \\
\mathbf{DA}_{\psi} \uparrow & & \uparrow \mathbf{DA}_{\hat{\psi}} \\
\mathbf{DA}_J & \xleftarrow{\mathbf{DA}_{\hat{\phi}}} & \mathbf{DA}_{I \cup J}
\end{array}
$$

For the two automata in Definition 1 we will have, in general, $\mathbf{DA}_{\phi}(\mathcal{M}_1) \neq \mathbf{DA}_{\psi}(\mathcal{M}_2)$, thus $\mathcal{M}_1$ and $\mathcal{M}_2$ can be not co-amalgamated.

The only relation between co-amalgamation and synchronization is that $\mathcal{M}_1 \parallel \mathcal{M}_2$ can be reconstructed by coamalgamation from its "components" $\mathbf{DA}_{\hat{\psi}}(\mathcal{M}_1 \parallel \mathcal{M}_2)$ and $\mathbf{DA}_{\hat{\phi}}(\mathcal{M}_1 \parallel \mathcal{M}_2)$. This, however, is true for any $(I \cup J)_{\rightarrow}$-coalgebra since the diagram on the right above is a pullback.

In the given coalgebraic setting we can even not relate $\mathcal{M}_1 \parallel \mathcal{M}_2$ with the original automata $\mathcal{M}_1$ and $\mathcal{M}_2$: The projection $\pi_1 : S_1 \times S_2 \to S_1$, for example, does not provide an $I_{\rightarrow}$-homomorphism from $\mathbf{DA}_{\hat{\psi}}(\mathcal{M}_1 \parallel \mathcal{M}_2)$ into $\mathcal{M}_1$ since definedness is preserved but not reflected,

$$
\begin{array}{ccc}
S_1 \times S_2 & \xrightarrow{\hat{\psi}; syn_{\mathcal{M}_1, \mathcal{M}_2}(\_)} & [I \twoheadrightarrow S_1 \times S_2] \\
\pi_1 \downarrow & & \downarrow \_; \pi_1 \\
S_1 & \xrightarrow{\beta_1} & [I \twoheadrightarrow S_1]
\end{array}
$$

i.e., we have only $\pi_1; \beta_1 \geq \hat{\psi}; syn_{\mathcal{M}_1, \mathcal{M}_2}(\_); \pi_1$, where $\geq$ reflects the corresponding partial order on $[I \twoheadrightarrow S_1]$.

An idea could be now to allow for those kind of "weak homomorphims" and to try to characterize $\mathcal{M}_1 \parallel \mathcal{M}_2$ by a minimality property: There exists for any $(I \cup J)_{\rightarrow}$-coalgebra $\mathcal{M}$, any weak $I_{\rightarrow}$-homomorphism $f : \mathbf{DA}_{\hat{\psi}}(\mathcal{M}) \to \mathcal{M}_1$, and any weak $J_{\rightarrow}$-homomorphism $g : \mathbf{DA}_{\hat{\phi}}(\mathcal{M}) \to \mathcal{M}_2$ a unique weak $(I \cup J)_{\rightarrow}$-homomorphism $h : \mathcal{M} \to \mathcal{M}_1 \parallel \mathcal{M}_2$ such that $\mathbf{DA}_{\hat{\psi}}(h); \pi_1 = f$ and

$\mathbf{DA}_{\hat{\phi}}(h); \pi_2 = g$. Even this characterization, however, fails and the problem is not the uniqueness but the existence of h.

6 Synchronization Algebraically

In the last section we have seen that synchronization can be not described in terms of co-amalgamation, i.e., within the coalgebraic setting. Therefore we will try to find in this section a characterization of synchronization within a partial algebraic setting. Firstly, we have in this setting weak homomorphisms available thus the problem concerning the projections can be resolved. Secondly, it allows to fix the intuition that $\mathcal{M}_1 \parallel \mathcal{M}_2$ is a subautomaton of a product automaton. In other formalisms as CommUnity [4], for example, synchronization is described by a pullback (on the level of sets of actions). Thus we will look for a pullback characterization of synchronization.

The coalgebraic setting in the first sections has been an "indexed setting" since we have considered, in accordance with many other approaches, sets of actions as signatures. To obtain the intended universal characterization of synchronization we have to work, however, in a "flat setting", i.e., we have to consider sets of actions not as signatures but only as one of the two carrier sets of an automaton. The relation between the "indexed setting" and the "flat setting" will be clarified by a general result presented in the next section.

As already pointed out, a deterministic partial automaton $\mathcal{A}$ without output can be seen as a *partial AUT-algebra* where AUT is a specification (signature) given by two sort symbols In and St and one operation symbol t with arity $t : St\,In \to St$. A *(weak) AUT-homomorphism* $f : \mathcal{A} \to \mathcal{B}$ between two AUT-algebras $\mathcal{A}$ and $\mathcal{B}$ is given by two maps $f(In) : \mathcal{A}(In) \to \mathcal{B}(In)$ and $f(St) : \mathcal{A}(St) \to \mathcal{B}(St)$ such that $f(St) \times f(In)(\mathrm{dom}\mathcal{A}(t)) \subseteq \mathrm{dom}\mathcal{B}(t)$ and $f(St)(\mathcal{A}(t)(s,i)) = \mathcal{B}(t)(f(St)(s), f(In)(i))$ for all $(s,i) \in \mathrm{dom}\mathcal{A}(t)$. f is said to be *closed* if f also reflects definedness, i.e., if we have $(s,i) \in \mathrm{dom}\mathcal{A}(t)$ iff $(f(St)(s), f(In)(i)) \in \mathrm{dom}\mathcal{B}(t)$ for all $(s,i) \in \mathcal{A}(St) \times \mathcal{A}(In)$ (see [2,9,12]). By $\mathbf{PAlg}(AUT)$ $(\mathbf{PAlg}^{cl}(AUT))$ we denote the category of all partial AUT-algebras and all (closed) AUT-homomorphisms.

A closer look at the obvious uncurried algebraic version of synchronization makes apparent that we are not synchronizing simple automata but automata with "idle actions". That is, we have to consider a further specification $AUT_\perp$ that extends AUT by a constant symbol $e : \to In$ and by an existence equation $(s : St, t(s,e) \overset{e}{=} s)$ forcing the idle action to be always defined and not to change the state of the automaton.

We have now a free functor $F : \mathbf{PAlg}(AUT) \to \mathbf{PAlg}(AUT_\perp)$ and instead of the two original AUT-algebras $\mathcal{A}$ and $\mathcal{B}$ the corresponding two extended AUT-algebras $\mathcal{A}_\perp \overset{def}{=} U(F(\mathcal{A}))$ and $\mathcal{B}_\perp \overset{def}{=} U(F(\mathcal{B}))$ are synchronized, where $U : \mathbf{PAlg}(AUT_\perp) \to \mathbf{PAlg}(AUT)$ is the forgetful functor that forgets that the idle action was designated as a constant. In such a way, the extended AUT-algebra $\mathcal{A}_\perp$ will be given by

- $\mathcal{A}_\perp(St) \overset{def}{=} \mathcal{A}(St)$, $\mathcal{A}_\perp(In) \overset{def}{=} \mathcal{A}(In) \cup \{\perp_\mathcal{A}\}$,
- $\mathrm{dom}\mathcal{A}_\perp(t) \overset{def}{=} \mathrm{dom}\mathcal{A}(t) \cup \mathcal{A}(St) \times \{\perp_\mathcal{A}\}$ and for all $(s,i) \in \mathrm{dom}\mathcal{A}_\perp(t)$ we have

$$\mathcal{A}_\perp(t)(s,i) \overset{def}{=} \begin{cases} \mathcal{A}(t)(s,i), (s,i) \in \mathrm{dom}\mathcal{A}(t) \\ (s,i) \quad, i = \perp_\mathcal{A} \end{cases}$$

All limits in $\mathbf{PAlg}(AUT)$ exist and are obtained by constructing the corresponding limits in $\mathbf{Set}$. The product $\mathcal{A}_\perp \times \mathcal{B}_\perp$ of $\mathcal{A}_\perp$ and $\mathcal{B}_\perp$ is given by

- $\mathcal{A}_\perp \times \mathcal{B}_\perp(St) \overset{def}{=} \mathcal{A}(St) \times \mathcal{B}(St)$, $\mathcal{A}_\perp \times \mathcal{B}_\perp(In) \overset{def}{=} \mathcal{A}_\perp(In) \times \mathcal{B}_\perp(In)$,
- $\mathrm{dom}\mathcal{A}_\perp \times \mathcal{B}_\perp(t) \overset{def}{=} \mathrm{dom}\mathcal{A}_\perp(t) \times \mathrm{dom}\mathcal{B}_\perp(t) = (\mathrm{dom}\mathcal{A}(t) \cup \mathcal{A}(St) \times \{\perp_\mathcal{A}\}) \times (\mathrm{dom}\mathcal{B}(t) \cup \mathcal{B}(St) \times \{\perp_\mathcal{B}\})$ and for all $((s,s'),(i,i')) \in \mathrm{dom}\mathcal{A}_\perp \times \mathcal{B}_\perp(t)$ we have

$$\mathcal{A}_\perp \times \mathcal{B}_\perp(t)((s,s'),(i,i'))$$
$$\overset{def}{=} \begin{cases} (\mathcal{A}(t)(s,i),s') & , (s,i) \in \mathrm{dom}\mathcal{A}(t), i' = \perp_\mathcal{B} \\ (\mathcal{A}(t)(s,i),\mathcal{B}(t)(s',i')), (s,i) \in \mathrm{dom}\mathcal{A}(t), (s',i') \in \mathrm{dom}\mathcal{B}(t) \\ (s,\mathcal{B}(t)(s',i')) & , i = \perp_\mathcal{A}, (s',i') \in \mathrm{dom}\mathcal{B}(t) \\ (s,s') & , i = \perp_\mathcal{A}, i' = \perp_\mathcal{B} \end{cases}$$

The definition of $\mathcal{A}_\perp \times \mathcal{B}_\perp$ ensures that the projections $\pi_1(St) : \mathcal{A}(St) \times \mathcal{B}(St) \to \mathcal{A}(St)$, $\pi_1(In) : \mathcal{A}_\perp(In) \times \mathcal{B}_\perp(In) \to \mathcal{A}_\perp(St)$ and $\pi_2(St) : \mathcal{A}(St) \times \mathcal{B}(St) \to \mathcal{B}(St)$, $\pi_2(In) : \mathcal{A}_\perp(In) \times \mathcal{B}_\perp(In) \to \mathcal{B}_\perp(St)$ provide AUT-homomorphisms $\pi_1 : \mathcal{A}_\perp \times \mathcal{B}_\perp \to \mathcal{A}_\perp$ and $\pi_2 : \mathcal{A}_\perp \times \mathcal{B}_\perp \to \mathcal{B}_\perp$, respectively.

The (uncurried version of) the synchronized automaton $\mathcal{A} \parallel \mathcal{B}$ has the same carrier as $\mathcal{A}_\perp \times \mathcal{B}_\perp$. Moreover, the map $em : \mathcal{A}(In) \cup \mathcal{B}(In) \to \mathcal{A}_\perp(In) \times \mathcal{B}_\perp(In)$ with

$$em(x) \overset{def}{=} \begin{cases} (x,\perp_\mathcal{B}), x \in \mathcal{A}(In) \setminus \mathcal{B}(In) \\ (x,x) \quad, x \in \mathcal{A}(In) \cap \mathcal{B}(In) \\ (\perp_\mathcal{A},x), x \in \mathcal{B}(In) \setminus \mathcal{A}(In) \end{cases}$$

is injective and a comparison of the definitions above with Definition 1 makes immediately evident that the two maps $m(St) \overset{def}{=} id_{\mathcal{A}(St) \times \mathcal{B}(St)} : \mathcal{A} \parallel \mathcal{B}(St) \to \mathcal{A}_\perp \times \mathcal{B}_\perp(St)$ and $m(In) \overset{def}{=} em : \mathcal{A} \parallel \mathcal{B}(In) \to \mathcal{A}_\perp \times \mathcal{B}_\perp(In)$ define an AUT-homomorphism $m : \mathcal{A} \parallel \mathcal{B} \to \mathcal{A}_\perp \times \mathcal{B}_\perp$. That is, $\mathcal{A} \parallel \mathcal{B}$ can be essentially considered as an AUT-subalgebra of $\mathcal{A}_\perp \times \mathcal{B}_\perp$.

In view of pullbacks em can be described as the coequalizer of $\pi_1(In); \phi$ and $\pi_2(In); \psi$ in the following diagram where

$$\phi(x) \stackrel{def}{=} \begin{cases} \bot_L, x \in \mathcal{A}(In) \setminus \mathcal{B}(In) \\ x \; , x \in \mathcal{A}(In) \cap \mathcal{B}(In) \\ \bot_R, x = \bot_{\mathcal{A}} \end{cases} \quad \text{and} \quad \psi(x) \stackrel{def}{=} \begin{cases} \bot_R, x \in \mathcal{B}(In) \setminus \mathcal{A}(In) \\ x \; , x \in \mathcal{A}(In) \cap \mathcal{B}(In) \\ \bot_L, x = \bot_{\mathcal{B}} \end{cases}$$

thus the outer square in the diagram with $\hat{\psi} = em; \pi_1(In)$ and $\hat{\phi} = em; \pi_2(In)$ becomes a pullback diagram.

Note that the idle actions in $\mathcal{A}_\bot$ and $\mathcal{B}_\bot$ don't play the same rôle w.r.t. synchronization as they do it, for example, in CommUnity [4]. The crucial point is that $\phi(\bot_{\mathcal{A}}) \neq \psi(\bot_{\mathcal{B}})$, thus an action a in $\mathcal{A}_\bot(In)$ with $\phi(a) = \phi(\bot_{\mathcal{A}})$ can not be combined freely with an arbitrary action b in $\mathcal{B}_\bot(In)$ with $\psi(b) = \psi(\bot_{\mathcal{B}})$. On the contrary, all actions have to synchronize. The idle action $\bot_{\mathcal{A}}$ is forced to synchronize with an action in $\mathcal{B}_\bot(In) \setminus \mathcal{A}_\bot(In)$, and $\bot_{\mathcal{B}}$ is correspondingly forced to synchronize with an action in $\mathcal{A}_\bot(In) \setminus \mathcal{B}_\bot(In)$. And this constraint can be seen as the model-theoretic essence of lock-step synchronization in CSP.

To complete the pullback picture we have to extend $\mathcal{A}(In) \cap \mathcal{B}(In) \cup \{\bot_L, \bot_R\}$ to an AUT-algebra such that ϕ and ψ, respectively, can be extended to AUT-homomorphisms: We simply define a one-state total AUT-algebra $\mathcal{L}$ by

- $\mathcal{L}(St) \stackrel{def}{=} \{*\}, \quad \mathcal{L}(In) \stackrel{def}{=} \mathcal{A}(In) \cap \mathcal{B}(In) \cup \{\bot_L, \bot_R\}$,
- $\mathsf{dom}\mathcal{L}(t) \stackrel{def}{=} \mathcal{L}(St) \times \mathcal{L}(In)$ and $\mathcal{L}(t)(*, i) \stackrel{def}{=} *$ for all $i \in \mathcal{L}(In)$.

and obtain AUT-homomorphisms $f : \mathcal{A}_\bot \to \mathcal{L}$ and $g : \mathcal{B}_\bot \to \mathcal{L}$, respectively, with $f(In) \stackrel{def}{=} \phi$, $g(In) \stackrel{def}{=} \psi$ and with $f(St) : \mathcal{A}_\bot(St) \to \{*\}$, $g(St) : \mathcal{B}_\bot(St) \to \{*\}$ the obvious total constant functions. Finally, we obtain, in such a way, the intended pullback diagram in $\mathbf{PAlg}(AUT)$ since $m(St) = id_{\mathcal{A}(St) \times \mathcal{B}(St)} : \mathcal{A} \parallel \mathcal{B}(St) \to \mathcal{A}_\bot \times \mathcal{B}_\bot(St)$ is trivially the equalizer of $\pi_1(St); f(St)$ and $\pi_2(St); g(St)$

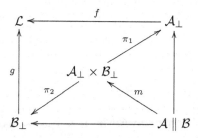

Note that the above discussion concerning lock-step synchronization is reflected by the fact that we construct a pullback in $\mathbf{PAlg}(AUT)$ but not in $\mathbf{PAlg}(AUT_\bot)$.

7   Signatures vs. Parameter Algebras

In section 4 we have insisted on abstract signature morphisms, i.e., on natural transformations, since we have been interested in dualizing amalgamation. In this section we want to relate the "indexed" coalgebraic setting and the "flat" algebraic setting, thus it is more appropriated now to consider instead the underlying maps between input sets. That is, we consider here the contravariant model functor $\mathbf{DA} : \mathbf{Set}^{op} \to \mathbf{Cat}$ given by the assigments $I \mapsto \mathbf{DA}_I$ and $\phi \mapsto \mathbf{DA}_\phi$.

Often the transition from an "indexed" to a "flat" setting can be described by the so-called Grothendieck construction [1]. This construction applied to $\mathbf{DA} :$ $\mathbf{Set}^{op} \to \mathbf{Cat}$ provides a category $\mathbf{Fl}(\mathbf{DA})$ defined as follows:

- An object of $\mathbf{Fl}(\mathbf{DA})$ is a pair $(I, (A, \alpha))$ where I is an input set and (A, α) is an $I_\to$-coalgebra.
- An arrow $(\phi, f) : (I, (A, \alpha)) \to (J, (B, \beta))$ in $\mathbf{Fl}(\mathbf{DA})$ has $\phi : I \to J$ a map and $f : (A, \alpha) \to (B, \phi; \beta(\_))$ is a $I_\to$-homomorphism.
- If $(\phi, f) : (I, (A, \alpha)) \to (J, (B, \beta))$ and $(\psi, g) : (J, (B, \beta)) \to (K, (C, \gamma))$ then the composition $(\phi, f); (\psi, g) : (I, (A, \alpha)) \to (K, (C, \gamma))$ is defined by

$$(\phi, f); (\psi, g) \stackrel{def}{=} (\phi; \psi, f; g)$$

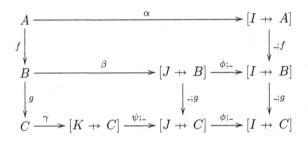

The construction induces a "projection" functor $P : \mathbf{Fl}(\mathbf{DA}) \to \mathbf{Set}$ given by $P(I, (A, \alpha)) \stackrel{def}{=} I$ and $P(\phi, f) \stackrel{def}{=} \phi$, and a general property of the Grothendieck construction is that this functor is a *split fibration* [1].

Every pair $(I, (A, \alpha))$ can be equivalently represented as an AUT-algebra $\mathcal{A}$ with $\mathcal{A}(In) \stackrel{def}{=} I$, $\mathcal{A}(St) \stackrel{def}{=} A$ and $\mathcal{A}(t) : \mathcal{A}(St) \times \mathcal{A}(In) \to \mathcal{A}(St)$ and with a transition function defined by uncurrying α, i.e., $(a, i) \in \mathrm{dom}\mathcal{A}(t)$ iff $i \in \mathrm{dom}(\alpha(a))$ and $\mathcal{A}(t)(a, i) \stackrel{def}{=} \alpha(a)(i)$ for all $(a, i) \in \mathrm{dom}\mathcal{A}(t)$. Moreover, any arrow $(\phi, f) : (I, (A, \alpha)) \to (J, (B, \beta))$ can be represented by a closed AUT-homomorphism $h : \mathcal{A} \to \mathcal{B}$ between the corresponding AUT-algebras $\mathcal{A}$ and $\mathcal{B}$ defined by $h(In) \stackrel{def}{=} \phi$ and $h(St) \stackrel{def}{=} f$. It is easy to see that this uncurrying provides

Proposition 1. *The categories and* $\mathbf{Fl}(\mathbf{DA})$ *and* $\mathbf{PAlg}^{cl}(AUT)$ *are isomorphic.*

By Proposition 1 we have now a transition from the "indexed"'coalgebraic setting into the (closed) "flat" algebraic setting. We have, however, even more. The "flat" setting turns out to be actually a "fibred" setting since we have according to the remarks above and Proposition 1 a split fibration $P : \mathbf{PAlg}^{cl}(AUT) \to \mathbf{Set}$ and the fibration can be transformed into the indexed category $\mathbf{DA} : \mathbf{Set}^{op} \to \mathbf{Cat}$ and vice versa.

If we consider now **Set** as the category of all IN-algebras where IN is the subsignature of AUT just given by the single sort In then we can formulate the question: Under which conditions do we have the free choice to consider a "structure" as a "signature" or as a "parameter algebra"?

Having a closer "algebraic" look at the definition of the coforgetful functors $\mathbf{DA}_\phi$ we observe that the "new operation" in AUT has as result sort also a "new sort" and not an old "parameter sort". This observation is the key for answering our question:

Theorem 2. *Let be given signatures $\Sigma_1 = (S_1, OP_1)$, $\Sigma_2 = (S_2, OP_2)$ and an embedding $\varphi : \Sigma_1 \to \Sigma_2$ with $s \in S_2 \setminus \varphi(S_1)$ for all op $: s_1 \ldots s_n \to s$ in $OP_2 \setminus \varphi(OP_1)$. Then the forgetful functor $U : \mathbf{PAlg}(\Sigma_2) \to \mathbf{PAlg}(\Sigma_1)$ is a split fibration.*

Proof: We assume w.l.o.g. $\Sigma_1 \subseteq \Sigma_2$. Recall that for any Σ_2-algebra $\mathcal{A}$ the forgetful image $U(\mathcal{A})$ is defined by $U(\mathcal{A})(s) = \mathcal{A}(s)$ for all $s \in S_1$ and by $U(\mathcal{A})(op) = \mathcal{A}(op)$ for all $op \in OP_1$. Moreover, for any Σ_2-homomorphism $g : \mathcal{A} \to \mathcal{B}$ the Σ_1-homomorphism $U(g) : U(\mathcal{A}) \to (\mathcal{B})$ is defined by $U(g)(s) = g(s)$ for all $s \in S_1$.

First, we have to define for any Σ_1-homomorphism $f : \mathcal{A}_1 \to \mathcal{A}_2$ and any Σ_2-algebra $\mathcal{B}$ with $U(\mathcal{B}) = \mathcal{A}_2$ a *cartesian arrow* $\gamma(f, \mathcal{B}) : \mathcal{C}(f, \mathcal{B}) \to \mathcal{B}$: For any $s \in S_2$ we set

$$\mathcal{C}(f, \mathcal{B})(s) \stackrel{def}{=} \begin{cases} \mathcal{B}(s), s \in S_2 \setminus S_1 \\ \mathcal{A}_1(s), s \in S_1 \end{cases}$$

and

$$\gamma(f, \mathcal{B})(s) \stackrel{def}{=} \begin{cases} id_{\mathcal{B}(s)}, s \in S_2 \setminus S_1 \\ f(s), s \in S_1 \end{cases}$$

Further, we set for any $op : s_1 \ldots s_n \to s$ in OP_2

$$\mathcal{C}(f, \mathcal{B})(op) \stackrel{def}{=} \begin{cases} (\gamma(f, \mathcal{B})(s_1) \times \ldots \times \gamma(f, \mathcal{B})(s_n)); \mathcal{B}(op), op \in OP_2 \setminus OP_1 \\ \mathcal{A}_1(op) \qquad\qquad\qquad\qquad\qquad\qquad , op \in OP_1 \end{cases}$$

In case $op \in OP_1$, the homomorphism property of $\mathcal{C}(f, \mathcal{B})(op)$ w.r.t. $\gamma(f, \mathcal{B})$ is exactly the homomorphism property of $\mathcal{A}_1(op)$ w.r.t. f, and, in case $op \in OP_2 \setminus OP_1$ the homomorphism property of $\mathcal{C}(f, \mathcal{B})(op)$ w.r.t. $\gamma(f, \mathcal{B})$ is ensured by the definition of $\mathcal{C}(f, \mathcal{B})(op)$ and since we have $\gamma(f, \mathcal{B})(s) = id_{\mathcal{B}(s)}$ by definition and by assumption.

The first *cartesian property* $U(\gamma(f, \mathcal{B})) = f$ and thus $U(\mathcal{C}(f, \mathcal{B})) = \mathcal{A}_1$ is directly ensured by definition. Moreover, also the *splitting conditions*

- $\gamma(id_{U(\mathcal{B})}, \mathcal{B}) = id_{\mathcal{B}}$ for all Σ_2-algebras $\mathcal{B}$ and
- $\gamma(f, \mathcal{C}(g, \mathcal{B}')); \gamma(g, \mathcal{B}') = \gamma(f; g, \mathcal{B}')$ for all Σ_1-homomorphisms $f : \mathcal{A}_1 \to \mathcal{A}_2$, $g : \mathcal{A}_2 \to \mathcal{A}_3$ and all Σ_2-algebras $\mathcal{B}'$ with $\mathcal{A}_3 = U(\mathcal{B}')$, $\mathcal{A}_2 = U(\mathcal{C}(g, \mathcal{B}'))$.

are immmediately ensured by definition. It remains to show the second cartesian property: For any Σ_2-homomorphism $v : \mathcal{C} \to \mathcal{B}$ and any Σ_1-homomorphism

$h : U(\mathcal{C}) \to \mathcal{A}_1$ for which $h; f = U(v)$ there is a unique Σ_2-homomorphism $\varepsilon(h, v) : \mathcal{C} \to \mathcal{C}(f, \mathcal{B})$ such that $U(\varepsilon(h, v)) = h$ and $\varepsilon(h, v); \gamma(f, \mathcal{B}) = v$.

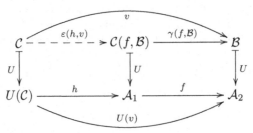

For any $s \in S_2$ we set

$$\varepsilon(h, v)(s) \stackrel{def}{=} \begin{cases} v(s), s \in S_2 \setminus S_1 \\ h(s), s \in S_1 \end{cases}$$

thus $U(\varepsilon(h, v)) = h$ and $\varepsilon(h, v); \gamma(f, \mathcal{B}) = v$ are immediately ensured by the definitions of $\varepsilon(h, v)$, $\gamma(f, \mathcal{B})$, U and by the assumption $h; f = U(v)$. Also the homomorphism property of $\varepsilon(h, v)$ can be shown straightforwardly.

The uniqueness of $\varepsilon(h, v)$ is forced, in case $s \in S_1$, by the requirement $U(\varepsilon(h, v)) = h$ and, in case $s \in S_2 \setminus S_1$, by the requirement $\varepsilon(h, v); \gamma(f, \mathcal{B}) = v$ and since we have $\gamma(f, \mathcal{B})(s) = id_{\mathcal{B}(s)}$ in this case.   □

Any split fibration gives rise to a contravariant functor into the category **Cat** [1] thus Theorem 2 provides

Corollary 1. *The split fibration* $U : \mathbf{PAlg}(\Sigma_2) \to \mathbf{PAlg}(\Sigma_1)$ *gives rise to a contravariant functor* $mod : \mathbf{PAlg}(\Sigma_1)^{op} \to \mathbf{Cat}$:

- For a Σ_1-algebra $\mathcal{A}$ $mod(\mathcal{A})$ is the *fiber over* $\mathcal{A}$, i.e., $mod(\mathcal{A})$ has as objects all Σ_2-algebras $\mathcal{B}$ with $U(\mathcal{B}) = \mathcal{A}$ and as morphisms all Σ_2-homomorphisms $g : \mathcal{B} \to \mathcal{B}'$ with $U(g) = id_{\mathcal{A}}$.
- For a Σ_1-homomorphism $f : \mathcal{A}_1 \to \mathcal{A}_2$ and a Σ_2-algebra in $\mathcal{B}$ in $mod(\mathcal{A}_2)$ we have $mod(f)(\mathcal{B}) = \mathcal{C}(g, \mathcal{B})$.
- For a Σ_1-homomorphism $f : \mathcal{A}_1 \to \mathcal{A}_2$ and a Σ_2-homomorphism $g : \mathcal{B}_1 \to \mathcal{B}_2$ in $mod(\mathcal{A}_2)$ we have $mod(f)(g) = \varepsilon(id_{\mathcal{A}_1}, \gamma(f, \mathcal{B}_1); g)$.

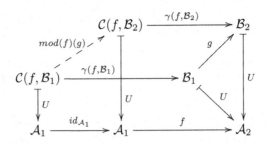

That these assignments preserve identities and composition is ensured by the splitting conditions.   □

The construction of a contravariant functor out of a split fibration and the Grothendieck construction are inverse to each other. This means, in our case, that the categories $\mathbf{Fl}(mod)$ and $\mathbf{PAlg}(\Sigma_2)$ are isomorphic. And, based on a variant of Theorem 2 for closed homomorphisms, that can be proved straightforwardly, this means that the contravariant functor $\mathbf{DA} : \mathbf{Set}^{op} \to \mathbf{Cat}$ and the contravariant functor $mod : \mathbf{Set}^{op} \to \mathbf{Cat}$ induced by the split fibration $U : \mathbf{PAlg}^{cl}(AUT) \to \mathbf{PAlg}^{cl}(IN)(\cong \mathbf{Set})$ are natural isomorphic.

8   Conclusions and Further Work

We have shown that there is co-amalgamation within the indexed coalgebraic reconstruction of deterministic CSP developed in [14]. It turned out that synchronization can not be characterized in terms of co-amalgamation.

We have been able, however, to characterize synchronization by a pullback property within the corresponding fibred algebraic setting. Analyzing and generalizing the transition between the indexed and the fibred setting we have shown that for a wide range of signature embeddings $\varphi : \Sigma_1 \to \Sigma_2$ the Σ_1-algebras, traditionally considered as parameter algebras, can be considered also as signatures, instead.

Our analysis makes apparent that there is a conceptual mismatch between process calculi and model theory. Concepts and constructions in process calculi are not directly related to model theoretic concepts and constructions and vice versa. To give a complete model theoretic account of constructions in deterministic CSP, we had to take different viewpoints and we had to move freely between them (compare also, for example, the analysis of *concealment* in [14]): We had to take into account different kinds of automata and the transitions between them. And we had to work as well in an indexed coalgebraic as in the corresponding fibred algebraic setting.

One conclusion could be that all the different viewpoints reflect different aspects of systems and that only a future structured synthesis of all the model theoretic viewpoints as well as the viewpoint of process calculi will provide an appropriate framework for system specifications.

There are at least three directions of further work:

- **Synchronization:** Firstly, we should analyse as well synchronization of non-deterministic processes in CSP as synchronization mechanisms in other process calculi. Secondly, the model theoretic description of synchronization by a special pullback opens the possibility to define other kinds of synchronization based on general pullbacks and to incorporate them into CSP.
- **Signatures vs. Parameter Algebras:** It is worth to look for other examples where Theorem 2 allows to consider a "structure" as a "signature" or as a "parameter algebra", respectively. And a more methodological discussion of this duality will be surely of interest.
- **Many-sorted Coalgebras:** The restriction on the signature extension in Theorem 2 allows also to transform, by currying, any Σ_2-algebra into a many-sorted coalgebra where the the sorts of these coalgebras are the sorts

in $S_2 \setminus S_1$. This transformation is a generalization the relation between $\mathbf{PAlg}^{cl}(AUT)$ and $\mathbf{DA}$. Co-amalgamation should be investigated for those many-sorted coalgebras. And, it would be nice to locate areas where many-sorted coalgebras are used or can be used with some benefit.

References

1. M. Barr and C. Wells. *Category Theory for Computing Science*. Series in Computer Science. Prentice Hall International, London, 1990.
2. P. Burmeister. *A Model Theoretic Oriented Approach to Partial Algebras*, volume 32 of *Mathematical Research — Mathematische Forschung*. Akademie-Verlag, Berlin, 1986.
3. H. Ehrig and B. Mahr. *Fundamentals of Algebraic Specification 1: Equations and Initial Semantics*, volume 6 of *EATCS Monographs on Theoretical Computer Science*. Springer, Berlin, 1985.
4. J. L. Fiadeiro. *Categories for Software Engineering*. Springer, Berlin, 2005.
5. J. A. Goguen and R. M. Burstall. Institutions: Abstract Model Theory for Specification and Programming. *Journals of the ACM*, 39(1):95–146, January 1992.
6. C.A.R. Hoare. *Communicating Sequential Processes*. Prentice-Hall, 1985.
7. B. Jacobs and J. Rutten. A Tutorial on (Co)Algebras and (Co)Induction. *Bulletin of EATCS*, 62:222–259, June 1997.
8. D. Pattinson. Translating Logics for Coalgebras. In *Recent Trends in Algebraic Development Techniques*. 16th International Workshop, WADT 2002, Springer Verlag, LNCS, 2002. To appear.
9. H. Reichel. *Initial Computability, Algebraic Specifications, and Partial Algebras*. Oxford University Press, 1987.
10. J.J.M.M. Rutten. Automata and coinduction (an exersice in coalgebra). Technical Report SEN-R9803, CWI, 1998.
11. J.J.M.M. Rutten. Universal coalgebra: A theory of systems. *TCS*, 249:3–80, 2000.
12. U. Wolter. An Algebraic Approach to Deduction in Equational Partial Horn Theories. *J. Inf. Process. Cybern. EIK*, 27(2):85–128, 1990.
13. U. Wolter. On Corelations, Cokernels, and Coequations. In H. Reichel, editor, *Third Workshop on Coalgebraic Methods in Computer Science (CMCS'2000), Berlin, Germany, Proceedings*, volume 33 of ENTCS, pages 347–366. Elsevier Science, 2000.
14. U. Wolter. CSP, Partial Automata, and Coalgebras. *TCS*, 280:3–34, 2002.
15. U. Wolter. Cofree Coalgebras for Signature Morphisms. In H.-J. Kreowski, editor, *Formal Methods (Ehrig Festschrift)*, pages 275–290. Springer, LNCS 3393, 2005.

Author Index

Lecture Notes in Computer Science

For information about Vols. 1–4352

please contact your bookseller or Springer

Vol. 4403: S. Obayashi, K. Deb, C. Poloni, T. Hiroyasu, T. Murata (Eds.), Evolutionary Multi-Criterion Optimization. XIX, 954 pages. 2007.

Vol. 4401: N. Guelfi, D. Buchs (Eds.), Rapid Integration of Software Engineering Techniques. IX, 177 pages. 2007.

Vol. 4400: J.F. Peters, A. Skowron, V.W. Marek, E. Orłowska, R. Słowiński, W. Ziarko (Eds.), Transactions on Rough Sets VII, Part II. X, 381 pages. 2007.

Vol. 4399: T. Kovacs, X. Llorà, K. Takadama, P.L. Lanzi, W. Stolzmann, S.W. Wilson (Eds.), Learning Classifier Systems. XII, 345 pages. 2007. (Sublibrary LNAI).

Vol. 4398: S. Marchand-Maillet, E. Bruno, A. Nürnberger, M. Detyniecki (Eds.), Adaptive Multimedia Retrieval: User, Context, and Feedback. XI, 269 pages. 2007.

Vol. 4397: C. Stephanidis, M. Pieper (Eds.), Universal Access in Ambient Intelligence Environments. XV, 467 pages. 2007.

Vol. 4396: J. García-Vidal, L. Cerdà-Alabern (Eds.), Wireless Systems and Mobility in Next Generation Internet. IX, 271 pages. 2007.

Vol. 4395: M. Daydé, J.M.L.M. Palma, Á.L.G.A. Coutinho, E. Pacitti, J.C. Lopes (Eds.), High Performance Computing for Computational Science - VECPAR 2006. XXIV, 721 pages. 2007.

Vol. 4394: A. Gelbukh (Ed.), Computational Linguistics and Intelligent Text Processing. XVI, 648 pages. 2007.

Vol. 4393: W. Thomas, P. Weil (Eds.), STACS 2007. XVIII, 708 pages. 2007.

Vol. 4392: S.P. Vadhan (Ed.), Theory of Cryptography. XI, 595 pages. 2007.

Vol. 4391: Y. Stylianou, M. Faundez-Zanuy, A. Esposito (Eds.), Progress in Nonlinear Speech Processing. XII, 269 pages. 2007.

Vol. 4390: S.O. Kuznetsov, S. Schmidt (Eds.), Formal Concept Analysis. X, 329 pages. 2007. (Sublibrary LNAI).

Vol. 4389: D. Weyns, H.V.D. Parunak, F. Michel (Eds.), Environments for Multi-Agent Systems III. X, 273 pages. 2007. (Sublibrary LNAI).

Vol. 4385: K. Coninx, K. Luyten, K.A. Schneider (Eds.), Task Models and Diagrams for Users Interface Design. XI, 355 pages. 2007.

Vol. 4384: T. Washio, K. Satoh, H. Takeda, A. Inokuchi (Eds.), New Frontiers in Artificial Intelligence. IX, 401 pages. 2007. (Sublibrary LNAI).

Vol. 4383: E. Bin, A. Ziv, S. Ur (Eds.), Hardware and Software, Verification and Testing. XII, 235 pages. 2007.

Vol. 4381: J. Akiyama, W.Y.C. Chen, M. Kano, X. Li, Q. Yu (Eds.), Discrete Geometry, Combinatorics and Graph Theory. XI, 289 pages. 2007.

Vol. 4380: S. Spaccapietra, P. Atzeni, F. Fages, M.-S. Hacid, M. Kifer, J. Mylopoulos, B. Pernici, P. Shvaiko, J. Trujillo, I. Zaihrayeu (Eds.), Journal on Data Semantics VIII. XV, 219 pages. 2007.

Vol. 4379: M. Südholt, C. Consel (Eds.), Object-Oriented Technology. VIII, 157 pages. 2007.

Vol. 4378: I. Virbitskaite, A. Voronkov (Eds.), Perspectives of Systems Informatics. XIV, 496 pages. 2007.

Vol. 4377: M. Abe (Ed.), Topics in Cryptology – CT-RSA 2007. XI, 403 pages. 2006.

Vol. 4376: E. Frachtenberg, U. Schwiegelshohn (Eds.), Job Scheduling Strategies for Parallel Processing. VII, 257 pages. 2007.

Vol. 4374: J.F. Peters, A. Skowron, I. Düntsch, J. Grzymała-Busse, E. Orłowska, L. Polkowski (Eds.), Transactions on Rough Sets VI, Part I. XII, 499 pages. 2007.

Vol. 4373: K. Langendoen, T. Voigt (Eds.), Wireless Sensor Networks. XIII, 358 pages. 2007.

Vol. 4372: M. Kaufmann, D. Wagner (Eds.), Graph Drawing. XIV, 454 pages. 2007.

Vol. 4371: K. Inoue, K. Satoh, F. Toni (Eds.), Computational Logic in Multi-Agent Systems. X, 315 pages. 2007. (Sublibrary LNAI).

Vol. 4370: P.P Lévy, B. Le Grand, F. Poulet, M. Soto, L. Darago, L. Toubiana, J.-F. Vibert (Eds.), Pixelization Paradigm. XV, 279 pages. 2007.

Vol. 4369: M. Umeda, A. Wolf, O. Bartenstein, U. Geske, D. Seipel, O. Takata (Eds.), Declarative Programming for Knowledge Management. X, 229 pages. 2006. (Sublibrary LNAI).

Vol. 4368: T. Erlebach, C. Kaklamanis (Eds.), Approximation and Online Algorithms. X, 345 pages. 2007.

Vol. 4367: K. De Bosschere, D. Kaeli, P. Stenström, D. Whalley, T. Ungerer (Eds.), High Performance Embedded Architectures and Compilers. XI, 307 pages. 2007.

Vol. 4366: K. Tuyls, R. Westra, Y. Saeys, A. Nowé (Eds.), Knowledge Discovery and Emergent Complexity in Bioinformatics. IX, 183 pages. 2007. (Sublibrary LNBI).

Vol. 4364: T. Kühne (Ed.), Models in Software Engineering. XI, 332 pages. 2007.

Vol. 4362: J. van Leeuwen, G.F. Italiano, W. van der Hoek, C. Meinel, H. Sack, F. Plášil (Eds.), SOFSEM 2007: Theory and Practice of Computer Science. XXI, 937 pages. 2007.

Vol. 4361: H.J. Hoogeboom, G. Păun, G. Rozenberg, A. Salomaa (Eds.), Membrane Computing. IX, 555 pages. 2006.

Vol. 4360: W. Dubitzky, A. Schuster, P.M.A. Sloot, M. Schroeder, M. Romberg (Eds.), Distributed, High-Performance and Grid Computing in Computational Biology. X, 192 pages. 2007. (Sublibrary LNBI).

Vol. 4358: R. Vidal, A. Heyden, Y. Ma (Eds.), Dynamical Vision. IX, 329 pages. 2007.

Vol. 4357: L. Buttyán, V. Gligor, D. Westhoff (Eds.), Security and Privacy in Ad-Hoc and Sensor Networks. X, 193 pages. 2006.

Vol. 4355: J. Julliand, O. Kouchnarenko (Eds.), B 2007: Formal Specification and Development in B. XIII, 293 pages. 2006.

Vol. 4354: M. Hanus (Ed.), Practical Aspects of Declarative Languages. X, 335 pages. 2006.

Vol. 4353: T. Schwentick, D. Suciu (Eds.), Database Theory – ICDT 2007. XI, 419 pages. 2006.